ICD-11 Made Easy

The Complete Transition Guide from ICD-10 with Mapping Tables and Practice Scenarios

I0039612

Ehren Vincent Houston

ISBN: 978-1-7642339-0-3
Isohan Publishing

Table of Contents

Chapter 1: Introduction to ICD-11 and Key Differences from ICD-10

The medical world stands at a crossroads. After decades of using ICD-10, healthcare professionals now face a new classification system that promises to revolutionize how we document, track, and understand disease. ICD-11 isn't just an update—it's a fundamental reimagining of medical classification for the digital age.

What is ICD-11?

The International Classification of Diseases, 11th Revision (ICD-11) represents the World Health Organization's most ambitious undertaking in medical classification history (1). Unlike its predecessors, ICD-11 was born digital, designed from the ground up to work seamlessly with electronic health records, artificial intelligence systems, and global health databases.

Think of ICD-10 as a paper atlas translated into digital format. Now imagine ICD-11 as Google Maps—interactive, constantly updated, and inherently digital. This isn't merely a new version; it's a complete paradigm shift in how we conceptualize and organize medical knowledge.

The system contains over 55,000 unique codes compared to ICD-10's 14,400, but the real innovation lies not in quantity but in flexibility (2). ICD-11 introduces post-coordination—the ability to combine codes to create precise clinical descriptions. This means clinicians can now accurately capture the full complexity of a patient's condition without being constrained by pre-defined code combinations.

Case Example 1: Complex Diabetes Documentation

Consider Maria, a 62-year-old patient with longstanding diabetes. Under ICD-10, her condition might be coded as E11.65 (Type 2 diabetes mellitus with hyperglycemia). But Maria's reality is far more complex. She has diabetic retinopathy in her left eye, stage 3 chronic kidney disease, and peripheral neuropathy affecting both feet.

In ICD-11, her condition can be precisely captured:

- Primary code: 5A11 (Type 2 diabetes mellitus)
- Post-coordinated with: 9B71.0Z (Diabetic retinopathy), affecting left eye
- Plus: 5A25.1 (Diabetic kidney disease, stage 3)
- Plus: 8C10.2 (Diabetic polyneuropathy)

This level of detail isn't just academic—it directly impacts treatment planning, research accuracy, and ultimately, patient outcomes.

Why the Transition Matters

Healthcare has transformed dramatically since ICD-10's development in the 1980s. We've discovered new diseases, developed novel treatments, and fundamentally changed our understanding of many conditions. ICD-10, despite updates, struggles to keep pace with modern medicine.

The COVID-19 pandemic exposed these limitations starkly. When the novel coronavirus emerged, the medical community had to wait for emergency codes to be created and distributed—a process that took precious weeks (3). ICD-11's flexible architecture would have allowed immediate, precise coding of this new disease and its variants.

Beyond crisis response, the transition matters for three critical reasons:

Scientific Accuracy: Medical knowledge doubles every 73 days (4). ICD-11 captures advances in genetics, immunology, and disease understanding that simply didn't exist when ICD-10 was created. Conditions like gaming disorder, now recognized as legitimate health concerns, have proper classification for the first time.

Global Standardization: Currently, different countries use modified versions of ICD-10—ICD-10-CM in the United States, ICD-10-CA in Canada, ICD-10-AM in Australia. These variations create barriers to international research and health data exchange. ICD-11 eliminates these regional modifications, creating a truly global language for health.

Digital Integration: Modern healthcare runs on data. Electronic health records, clinical decision support systems, and population health platforms all need standardized, machine-readable information. ICD-11's API-first design enables seamless integration with these digital tools.

Core Benefits for Healthcare Organizations

The transition to ICD-11 offers tangible benefits across multiple domains:

Clinical Excellence

Dr. James Chen, chief medical officer at a 500-bed hospital system, describes the clinical impact: "Post-coordination allows our physicians to document exactly what they see, not just the closest available code. This precision improves everything from treatment protocols to quality metrics" (5).

The system's expanded rare disease coverage—over 5,500 newly recognized conditions—means patients with uncommon diagnoses finally have proper classification. This visibility drives research funding, treatment development, and support services for previously overlooked populations.

Case Example 2: Rare Disease Recognition

Eight-year-old Ethan presented with developmental delays, distinctive facial features, and cardiac abnormalities. For years, his condition defied classification. ICD-10 offered only vague codes for "other specified syndromes."

With ICD-11's expanded genetic disorder classification, Ethan's condition—a specific microdeletion syndrome—has its own code. This proper classification connected his family to targeted research studies, specialized support groups, and emerging treatments specifically designed for his condition.

Operational Efficiency

While initial implementation requires investment, ICD-11's design ultimately streamlines operations:

- **Reduced Queries**: More specific coding means fewer clarification requests between coders and clinicians
- **Automated Validation**: Built-in rules prevent invalid code combinations
- **Simplified Maintenance**: Annual updates replace decade-long revision cycles
- **Enhanced Interoperability**: Native API support eliminates costly interface development

Financial Optimization

Accurate coding directly impacts revenue. ICD-11's specificity supports:

- More precise risk adjustment
- Better documentation of complexity
- Reduced claim denials
- Improved contract negotiations

A mid-sized health system in Scandinavia reported 12% fewer coding-related denials after ICD-11 implementation, translating to $3.2 million in recovered revenue (6).

Quick Comparison Chart: ICD-10 vs ICD-11

Understanding the fundamental differences between these systems helps organizations prepare for change:

Structural Differences

ICD-10 uses an alphanumeric system with codes ranging from three to seven characters. The first character is always a letter, followed by numbers, with an optional decimal point for subcategories. This rigid structure limits expansion—many sections are simply full.

ICD-11 employs a more flexible four-character base code (letter-letter-number-number) with unlimited extension possibilities. This design accommodates infinite granularity without running out of code space.

Functional Capabilities

Where ICD-10 offers static, pre-defined codes, ICD-11 provides dynamic coding through post-coordination. Think of it as the difference between choosing from a restaurant's fixed menu versus having a chef prepare exactly what you need.

ICD-10's paper-based heritage shows in its linear structure and limited cross-references. ICD-11's digital DNA enables:

- Multiple parent categories for conditions
- Rich semantic relationships
- Real-time updates
- Integrated decision support

Case Example 3: Mental Health Classification Evolution

Sarah, a 34-year-old teacher, seeks help for mood difficulties. Under ICD-10, her diagnosis might be coded as F32.1 (Moderate depressive episode). This tells part of her story but misses crucial context.

ICD-11's restructured mental health chapter allows coding that captures:

- 6A70.1 (Single episode depressive disorder, moderate severity)
- With anxious distress (post-coordinated)
- With melancholic features (post-coordinated)

- Seasonal pattern (extension code)

This comprehensive coding guides treatment selection, predicts medication response, and enables participation in targeted research protocols.

Implementation Considerations

The transition from ICD-10 to ICD-11 involves more than learning new codes. Organizations must consider:

Technology Infrastructure: ICD-11 requires modern systems capable of handling:

- UTF-8 character encoding for multilingual support
- RESTful API communications
- JSON/XML data formats
- Cloud-ready architectures

Workflow Redesign: Clinical documentation processes need updating to leverage post-coordination effectively. This means rethinking everything from assessment forms to discharge summaries.

Training Investment: While ICD-11's logical structure makes it easier to learn than ICD-10, the new features require comprehensive education. Studies show coders need approximately six months to reach full productivity (7).

Change Management: Success depends on engaging all stakeholders—from C-suite executives to front-line clinicians. Organizations that treat this as merely an IT project consistently struggle.

The transition to ICD-11 represents healthcare's biggest classification change in a generation. Organizations that understand not just what's changing but why it matters position themselves for success. The benefits—clinical precision, operational efficiency, and true global interoperability—far outweigh the implementation challenges.

Start preparing now. Assess your current systems, engage your stakeholders, and begin building the foundation for this transformation. The question isn't whether to adopt ICD-11, but how quickly and effectively you can harness its power to improve patient care.

Key Takeaways

- ICD-11 is a complete reimagining of medical classification, not just an update to ICD-10
- Post-coordination enables precise clinical documentation previously impossible
- The system includes 55,000+ codes and covers 5,500+ rare diseases
- Digital-first design supports seamless integration with modern health IT
- Benefits span clinical accuracy, operational efficiency, and financial optimization
- Success requires comprehensive planning across technology, workflow, and people
- Early preparation and stakeholder engagement drive smooth transitions

Chapter 2: History and Development of ICD-11

The story of ICD-11 begins not with committees or code structures, but with a fundamental question: How do we create a classification system that can keep pace with the exponential growth of medical knowledge? The answer took 12 years, involved thousands of experts, and revolutionized our approach to health information.

The Journey from 2007 to 2025

In 2007, the World Health Organization faced a dilemma. ICD-10, though widely adopted, was showing its age. Medical science had advanced dramatically since its 1990s development—genomics had decoded human DNA, immunotherapy was transforming cancer treatment, and digital health was emerging as a new frontier (8). The WHO made a bold decision: rather than patch ICD-10 again, they would reimagine classification from scratch.

The initial vision seemed straightforward—create an updated classification system. But early planning sessions revealed a more profound opportunity. Dr. Robert Jakob, then leading the WHO Classifications team, recalls the pivotal moment: "We realized we weren't just updating codes. We were building the foundation for how humanity would understand and share health information for decades to come" (9).

Case Example 1: The Genomics Challenge

The need for change became clear through cases like the Williams family. Three siblings, all diagnosed with what ICD-10 could only classify as "familial hypercholesterolemia," actually had three distinct genetic mutations requiring different treatments. Their geneticist, Dr. Patricia Moore, explains: "ICD-10 gave us one code for dozens of different conditions. We were practicing 21st-century medicine with 20th-century classification tools."

The ICD-11 development team recognized that genetics would reshape medicine. They designed a system flexible enough to accommodate discoveries not yet made, diseases not yet understood.

The Development Phases

2007-2010: Foundation Building

The WHO assembled an unprecedented team. Unlike previous revisions dominated by statisticians and nosologists, ICD-11's development included:

- Practicing clinicians from every specialty
- Software engineers and database architects
- Patient advocacy groups
- Health informatics experts
- Linguists and translators

This diversity proved crucial. When engineers proposed a purely hierarchical structure, clinicians explained why diseases rarely fit neat categories. When statisticians suggested simplified coding, practitioners demonstrated the need for clinical nuance.

The team developed core principles:

1. Digital-first design—no paper legacy constraints
2. Scientific currency—accommodate ongoing discoveries
3. Global applicability—work across all healthcare settings
4. User-centricity—designed for those who actually use it

2011-2014: Alpha and Beta Development

May 2011 marked a milestone—the first public alpha release. Unlike traditional WHO processes conducted behind closed doors, ICD-11 development embraced radical transparency. Anyone could view the evolving classification, suggest changes, and participate in discussions.

This openness generated both excitement and chaos. The beta platform received over 10,000 proposals in its first year (10). Suggestions ranged from minor terminology updates to fundamental structural changes. Managing this input required innovative approaches.

The WHO created a distributed review system. Topic Advisory Groups (TAGs) for each medical specialty evaluated proposals. The groups included fascinating dynamics—cardiologists debating philosophers about disease definitions, patient advocates challenging traditional medical hierarchies.

Case Example 2: Mental Health Revolution

The mental health TAG faced particular challenges. ICD-10's approach to psychiatric classification, largely unchanged since the 1970s, poorly reflected modern understanding. Dr. Geoffrey Reed, leading the mental health revision, describes the process: "We had psychiatrists, psychologists, social workers, and—crucially—people with lived experience of mental illness all at the table. Some discussions got heated, but that diversity led to breakthrough insights."

One breakthrough involved depression classification. ICD-10's rigid episode-based model didn't match clinical reality. Patients described mood disorders as dimensional, fluctuating experiences. The TAG developed a new approach allowing clinicians to code severity, features, and patterns—capturing the true complexity of mental health conditions.

2015-2018: Field Testing and Refinement

Theory met reality during field trials. The WHO conducted studies across diverse settings:

- Japanese rural clinics
- Brazilian emergency departments
- Canadian teaching hospitals
- Nigerian community health centers
- German specialty practices

Each setting revealed unique challenges. Japanese clinicians found certain mental health concepts didn't translate culturally. Brazilian emergency physicians needed faster coding workflows. Nigerian health workers required systems functioning with limited internet connectivity.

Case Example 3: Cultural Adaptation Without Compromise

Dr. Kenji Tanaka led field trials in rural Japan. His team discovered that ICD-11's depression categories, while clinically sound, used concepts without direct Japanese equivalents. Rather than force awkward translations, the team worked with linguists to develop culturally appropriate terms maintaining clinical accuracy.

"The beauty of ICD-11," Tanaka notes, "is that it achieves true multilingual support without sacrificing precision. Previous systems simply translated English concepts. ICD-11 builds cultural understanding into its foundation" (11).

Global Collaboration and Stakeholders

The scale of collaboration for ICD-11 was unprecedented in medical history. By 2018, the development process had involved:

- 300+ experts in Topic Advisory Groups
- 90+ countries providing feedback
- 15,000+ registered users on the development platform
- 100,000+ comments and proposals processed

But numbers tell only part of the story. The real innovation lay in how these diverse voices shaped the final product.

Breaking Down Silos

Traditional medical classification operated in specialty silos. Cardiologists classified heart diseases, neurologists handled brain conditions, with minimal cross-talk. ICD-11 forced integration.

Consider diabetes—traditionally an endocrine disorder. The development process revealed diabetes complications spanning every organ system. Rather than scatter diabetes-related codes across chapters, ICD-11 enables post-coordination linking diabetic complications to their source.

The Patient Voice

For the first time in ICD history, patient advocates held formal positions in development. Their influence appears throughout the system:

- Plain language synonyms for medical terms
- Codes for patient-reported symptoms
- Recognition of conditions long dismissed by medicine
- Quality of life considerations in classification

Mary Chen, representing chronic fatigue syndrome patients, fought for recognition of post-exertional malaise—a defining symptom previously uncoded. "Doctors couldn't bill for addressing our primary concern because it didn't exist in their classification system. ICD-11 changes that" (12).

Technology Partners

Unlike previous paper-based revisions, ICD-11 required deep technology expertise. The WHO partnered with:

- Stanford University's Center for Biomedical Informatics
- National Institutes of Health (U.S.)
- DIMDI (Germany's medical documentation institute)
- SNOMED International

These partnerships went beyond technical support. Technology experts fundamentally influenced the classification's architecture, ensuring compatibility with emerging health IT standards.

WHO's Vision for Digital Health Classification

The World Health Organization's vision for ICD-11 extended far beyond updating disease codes. They imagined a living system that would:

Evolve Continuously

Previous ICD versions remained static for decades. ICD-9 lasted 30 years; ICD-10 is approaching 35. During these periods, medical knowledge expanded exponentially while classification stood still.

ICD-11 introduces annual updates—not just corrections but substantive additions reflecting new discoveries. The infrastructure supports this through:

- Modular architecture allowing section updates without system-wide changes
- Version control tracking all modifications
- Automated testing ensuring changes don't break existing functions
- Transparent change proposals visible to all users

Enable Semantic Interoperability

Dr. Christopher Chute, who led ICD-11's informatics architecture, explains the semantic web vision: "We're not just assigning codes to diseases. We're creating a web of meaning that machines can navigate and understand" (13).

This semantic foundation enables:

- Automatic translation between classification systems
- Intelligent clinical decision support
- Sophisticated epidemiological analysis
- AI-powered diagnostic assistance

Support Global Health Equity

Traditional classifications reflected the diseases of wealthy nations. Tropical diseases, conditions of poverty, and traditional medicine practices received minimal attention. ICD-11 corrects this imbalance.

Chapter 26, covering traditional medicine, exemplifies this commitment. Rather than dismissing non-Western healing systems, ICD-11 provides structure for documenting traditional medicine conditions alongside conventional diagnoses. This enables:

- Research into traditional remedies
- Integration of healing systems
- Respect for cultural practices
- Evidence-based evaluation of all interventions

Timeline of Major Milestones

The path from concept to implementation spans nearly two decades:

2007: WHO initiates ICD-11 development

- Establishes revision steering group
- Defines digital-first principles
- Begins recruiting global experts

2009: Technical architecture defined

- Semantic web foundation established
- API specifications drafted
- Multilingual framework designed

2011: Alpha version released

- First public viewing of draft
- Collaborative platform launched
- Feedback mechanisms activated

2012: Beta platform goes live

- Full classification browsable online
- Proposal system implemented
- Field trial protocols developed

2015: Major structural decisions finalized

- 28-chapter framework confirmed
- Post-coordination rules established
- Extension code system designed

2016: Field trials commence

- Multi-country testing begins
- Clinical coding studies launched
- Workflow impact assessed

2018: Stable version released (June 18)

- Content frozen for review
- Final field trial results analyzed
- Implementation guidance drafted

2019: World Health Assembly endorsement (May 25)

- Member states formally approve
- Transition planning begins
- Early adopter programs launch

2022: Official implementation date (January 1)

- ICD-11 becomes WHO standard
- 35 countries report active use
- First annual update cycle

2024: Widespread adoption phase

- 80+ countries in various implementation stages
- Major EHR vendors achieve compliance

- Research databases begin migration

2025: Maturation and optimization

- 132 Member States engaged
- Artificial intelligence integration
- Second generation tools emerge

Each milestone represents thousands of decisions, millions of data points, and countless hours of dedication from global health professionals.

The development of ICD-11 stands as a testament to what's possible when the global health community collaborates with shared purpose. From its 2007 inception to today's ongoing implementation, ICD-11 represents not just a new classification system but a new paradigm for how we create and maintain global health standards.

The journey continues. As you implement ICD-11 in your organization, you're not just adopting a classification system—you're joining a global movement to improve how we understand, document, and ultimately treat human disease.

Key Takeaways

- ICD-11 development spanned 12 years (2007-2019) with unprecedented global collaboration
- Over 300 experts from 55 countries shaped the classification through transparent, inclusive processes
- Digital-first design enables continuous updates versus decades-long static versions
- Field testing across diverse global settings ensured cultural relevance and practical utility
- Patient advocates formally participated for the first time in ICD history
- The semantic web foundation supports AI integration and intelligent health systems

- Annual updates keep pace with medical advances, unlike previous static versions
- Implementation continues expanding with 132 Member States actively engaged by 2025

Chapter 3: Structure and Organization of ICD-11

Understanding ICD-11's architecture is like learning a new language—once you grasp the underlying logic, everything else falls into place. The system's structure reflects decades of lessons learned from previous classifications while anticipating future needs we can barely imagine today.

The 28-Chapter Framework

The expansion from ICD-10's 21 chapters to ICD-11's 28 represents more than numerical growth—it signals a fundamental reorganization of how we conceptualize disease. Each new chapter addresses gaps that frustrated clinicians for decades.

Chapters 1-11: The Traditional Foundation

The first eleven chapters maintain familiar territory with significant improvements:

Chapter 1: Certain infectious or parasitic diseases reorganizes infections by causative organism rather than body system. A patient with disseminated tuberculosis affecting lungs, bones, and kidneys no longer needs codes scattered across multiple chapters. One primary code with post-coordination captures the complete picture.

Chapter 4: Diseases of the immune system stands as entirely new, recognizing immunology's emergence as a distinct medical specialty. Dr. Sarah Martinez, an immunologist in Barcelona, describes the impact: "We finally have codes for primary immunodeficiencies, autoinflammatory syndromes, and immune dysregulation disorders. Previously, we squeezed these complex conditions into inappropriate categories or used vague 'other specified' codes" (14).

Case Example 1: Immune System Classification

Nine-year-old Lucas presented with recurrent infections, autoimmune manifestations, and inflammatory episodes. ICD-10 forced his physician to choose between infection codes, autoimmune codes, or vague immunodeficiency categories. None captured his reality—an autoinflammatory syndrome affecting multiple systems.

ICD-11's immune system chapter provides specific coding:

- Primary code: 4A60.Y (Other specified autoinflammatory disorder)
- Post-coordinated with manifestation codes
- Extension codes for genetic markers
- Severity qualifiers for disease activity

This precise coding connected Lucas to appropriate clinical trials and secured insurance coverage for targeted biological therapies previously denied as "experimental."

Chapters 12-25: Evolution and Revolution

The middle chapters blend evolution of existing categories with revolutionary approaches:

Chapter 7: Sleep-wake disorders emerges as its own entity, acknowledging sleep medicine's growth. Previously scattered across neurology and psychiatry chapters, sleep disorders now have coherent organization. The chapter includes:

- Circadian rhythm disorders
- Sleep-related breathing disorders
- Parasomnias
- Sleep-related movement disorders

Chapter 17: Conditions related to sexual health represents perhaps the most culturally progressive change. Rather than scattering sexual health across infectious diseases, psychiatry, and genitourinary chapters, ICD-11 creates a stigma-free classification space. This includes:

- Sexual dysfunctions
- Gender incongruence (removed from mental disorders)
- Sexual pain disorders

Dr. Michael Thompson, who served on the sexual health Topic Advisory Group, explains: "We spent years debating every term, every category. The goal was clinical utility without perpetuating stigma. Gender incongruence's placement in this chapter rather than mental disorders sends a powerful message about acceptance and appropriate care" (15).

Case Example 2: Gender-Affirming Care Documentation

Alex, a 26-year-old seeking gender-affirming care, previously received psychiatric diagnoses that implied mental illness. Insurance companies often denied treatments as "not medically necessary" based on mental health classifications.

ICD-11 codes Alex's situation as:

- HA60 (Gender incongruence of adolescence or adulthood)
- Located in Chapter 17, not mental disorders
- Linkable to specific treatment codes
- No psychiatric implications

This reclassification transformed access to care. Insurance coverage improved, stigma decreased, and healthcare providers could document treatment needs without pathologizing identity.

Supplementary Chapters: Breaking New Ground

Chapter 26: Traditional Medicine Conditions marks ICD's first formal recognition of non-Western healing systems. Developed with traditional medicine practitioners from China, India, Japan, and Korea, this optional module enables:

- Research into traditional treatments
- Integration of healing modalities

- Insurance coverage for traditional practices
- Quality monitoring across systems

The traditional medicine chapter uses pattern differentiation rather than disease classification, respecting fundamental differences in medical philosophy while enabling documentation within ICD-11's framework.

Chapter X: Extension Codes provides the vocabulary for post-coordination. Think of these as adjectives and adverbs that modify primary diagnosis nouns. Categories include:

- Severity scales
- Temporality markers
- Anatomical detail
- Causality indicators
- Activity descriptors

Case Example 3: Complex Injury Documentation

Maria, a construction worker, fell from scaffolding, sustaining multiple injuries. ICD-10 would require numerous codes with unclear relationships. ICD-11 tells her complete story:

Primary injury code: NA25.2 (Fracture of shaft of femur) Post-coordinated with:

- XK8G (Laterality: right side)
- XS25 (Mechanism: fall from scaffold)
- XR34 (Activity: construction work)
- XT67 (Place of occurrence: industrial site)
- XJ45 (Initial encounter)

This clustering provides epidemiological data about workplace injuries while ensuring Maria receives appropriate acute care and worker's compensation benefits.

Understanding Foundation Component vs Linearizations

ICD-11's architecture separates knowledge representation from practical use—a distinction crucial for understanding the system's power and flexibility.

The Foundation Component: The Knowledge Universe

Imagine a vast library containing everything we know about human disease. Books can appear on multiple shelves, cross-reference each other infinitely, and exist in multiple languages simultaneously. That's the Foundation Component—ICD-11's semantic knowledge base.

The Foundation contains:

- 85,000+ unique entities
- Multiple inheritance (diseases can have multiple parents)
- Rich semantic relationships
- Complete clinical descriptions
- Synonyms in multiple languages

Dr. James Park, a health informaticist, uses this analogy: "If linearizations are like curated museum exhibits, the Foundation is the museum's entire collection—including pieces in storage, items being restored, and artifacts not yet on display" (16).

Linearizations: Practical Implementations

While the Foundation holds all knowledge, practical use requires organized subsets—linearizations. Each linearization serves specific purposes:

ICD-11 for Mortality and Morbidity Statistics (ICD-11-MMS): The primary linearization for health statistics, containing approximately 35,000 codes organized hierarchically. This is what most users mean when they say "ICD-11."

Primary Care Linearization: A simplified subset for general practice, focusing on common conditions with less granular detail.

Specialty Linearizations: Cardiology might use a linearization with extensive cardiovascular detail while simplifying other chapters.

Research Linearizations: Custom configurations for specific studies or registries.

The relationship between Foundation and linearizations enables unprecedented flexibility. Organizations can create custom linearizations for specific needs while maintaining mappings to standard classifications.

Code Syntax and Format

ICD-11's code structure balances human readability with machine processing needs. Understanding the syntax unlocks the system's power.

Basic Code Structure

Every ICD-11 code follows the pattern:
[Letter][Letter][Number][Number].[Decimal]

Examples:

- 1A00 (Cholera)
- BA00 (Essential hypertension)
- 5A11 (Type 2 diabetes mellitus)

The structure provides several advantages:

- First character indicates chapter (1-9, A-Z)
- Four-character stem codes are memorizable
- Decimal extensions add specificity
- No use of letters I or O (preventing confusion with 1 and 0)

Post-Coordination Syntax

Post-coordination uses specific symbols to build complex codes:

Forward Slash (/): Links related conditions Example: 5A11/5A25.1 (Type 2 diabetes with diabetic kidney disease)

Ampersand (&): Adds multiple extension codes Example: NC72.0&XK8J&XS25 (Ankle fracture, right side, due to fall)

The Grammar Rules

Like any language, post-coordination follows grammar rules:

1. **Stem codes first**: Always start with the primary condition
2. **Hierarchical ordering**: More significant elements before less significant
3. **Logical grouping**: Anatomical details together, temporal aspects together
4. **Validation requirements**: Not all combinations are valid

The system includes built-in validation preventing nonsensical combinations. You cannot code pregnancy in males or fractures of organs that have no bones.

Navigating the Digital Architecture

ICD-11's digital architecture goes beyond putting codes online—it fundamentally reimagines how classification systems work in the digital age.

The API Ecosystem

Application Programming Interfaces (APIs) allow systems to communicate directly with ICD-11. Rather than downloading static files, electronic health records can:

- Query for codes in real-time
- Validate combinations instantly
- Access the latest updates automatically
- Retrieve multilingual content on demand

Dr. Chen Wei, implementing ICD-11 at a large hospital network, describes the impact: "Our EHR talks directly to ICD-11. When WHO updates a code, our system knows immediately. No more manual updates, no more version confusion" (17).

The Semantic Web Integration

ICD-11 uses semantic web technologies, making it machine-readable and intelligent:

URIs (Uniform Resource Identifiers): Every entity has a permanent, unique identifier Example: http://id.who.int/icd/entity/1435254666

RDF (Resource Description Framework): Represents relationships between concepts

- "Diabetic retinopathy" IS-A "retinal disorder"
- "Diabetic retinopathy" IS-CAUSED-BY "diabetes mellitus"
- "Diabetic retinopathy" CAN-LEAD-TO "blindness"

SPARQL Queries: Enables sophisticated searches "Find all conditions that can cause kidney failure and are also associated with hypertension"

This semantic foundation enables artificial intelligence applications. Machine learning algorithms can traverse the relationship network, understanding disease connections beyond simple hierarchies.

Multilingual Architecture

Unlike previous ICDs that started in English then translated, ICD-11 builds multilingual capability into its core:

- Concept definitions exist independently of language
- Each concept links to terms in multiple languages
- Translations maintain semantic meaning, not just literal words
- Cultural adaptations possible without breaking mappings

25

Currently supporting 14 languages with 20+ in development, the architecture ensures global usability without sacrificing local relevance.

Real-Time Validation and Intelligence

ICD-11's digital nature enables intelligent coding assistance:

Conflict Detection: System identifies impossible combinations Example: Pregnancy codes with male gender markers trigger warnings

Completeness Checking: Alerts when required detail is missing Example: Injury codes without external cause prompt for additional information

Clinical Decision Support: Suggests related codes based on patterns Example: Coding diabetes triggers prompts about common complications

Quality Metrics: Tracks coding patterns for improvement Example: Overuse of unspecified codes generates training recommendations

The Architecture's Promise

ICD-11's structure represents more than technical specifications—it embodies a vision for health information's future. The 28-chapter framework provides room for medical knowledge to grow. The Foundation/linearization split enables both standardization and customization. The digital architecture ensures the classification evolves with technology rather than despite it.

As you work with ICD-11, appreciate how each structural decision supports better patient care. The system's complexity serves a purpose: capturing the full richness of human health and disease in ways that improve treatment, advance research, and ultimately save lives.

Key Takeaways

- ICD-11 expands to 28 chapters, adding crucial categories like immune system disorders and sleep-wake disorders
- New chapters address modern health needs, including sexual health conditions separate from mental disorders
- The Foundation Component contains 85,000+ entities with multiple relationships, while linearizations provide practical subsets
- Code syntax uses Letter-Letter-Number-Number format with post-coordination symbols (/ and &) for complex descriptions
- Digital architecture includes APIs, semantic web integration, and real-time validation
- Multilingual support is built into the core structure, not added through translation
- Extension codes (Chapter X) enable precise clinical detail through post-coordination
- Traditional medicine (Chapter 26) receives formal recognition for the first time in ICD history

Chapter 4: New Features and Improvements in ICD-11

The leap from ICD-10 to ICD-11 resembles the shift from film to digital photography. While both capture images, digital technology enables editing, sharing, and applications impossible with film. Similarly, ICD-11's innovations transform not just how we code diseases, but what we can do with health information.

Post-coordination and Cluster Coding

Post-coordination stands as ICD-11's most transformative feature. Rather than searching for pre-combined codes that never quite fit, clinicians can now build precise descriptions by combining elements—like constructing sentences from words rather than choosing from a phrase book.

The Mechanics of Post-coordination

Traditional ICD-10 coding resembles ordering from a fixed menu. You want chicken with vegetables and rice, but the menu only offers chicken with potatoes or fish with rice. You compromise, choosing the closest option.

ICD-11 works like a well-stocked kitchen. You select your protein (stem code), then add vegetables (extension codes), seasonings (severity qualifiers), and cooking method (temporal markers). The result precisely matches what you need.

The system uses two primary operators:

- **Forward slash (/)** connects related clinical concepts
- **Ampersand (&)** adds descriptive extensions

These simple symbols enable infinite combinations while maintaining logical structure.

Case Example 1: Oncology Precision

Dr. Patricia Kim treats a breast cancer patient with a complex presentation. Under ICD-10, she faces multiple inadequate options:

- C50.911 (Malignant neoplasm of unspecified site of right female breast)
- No code for specific histology
- Separate codes for lymph node involvement
- Additional codes for hormone receptor status

With ICD-11, she builds a complete picture:

```
2C61.1 (Invasive carcinoma of breast)
&XH3Y73 (Histology: Invasive ductal carcinoma)
&XK9J (Laterality: Right)
&XA9752 (Estrogen receptor positive)
&XA9754 (HER2 negative)
&2D11 (Regional lymph node metastasis)
```

This cluster captures everything needed for treatment planning, research enrollment, and quality reporting in a single, coherent expression (18).

Validation and Intelligence

Post-coordination includes built-in guardrails preventing illogical combinations. The system understands:

- Anatomical possibilities (you cannot code a fracture of the liver)
- Physiological constraints (males cannot have ovarian conditions)
- Temporal logic (chronic conditions cannot have acute onsets)

These rules operate silently, guiding users toward valid combinations without restricting legitimate clinical documentation.

Clinical Impact of Clustering

Cluster coding transforms several aspects of healthcare delivery:

Treatment Precision: Oncologists report that detailed tumor characterization through post-coordination directly influences chemotherapy selection. Dr. Robert Chen notes: "When I can code exact mutations and markers, clinical decision support systems provide targeted therapy recommendations. It's personalized medicine in action" (19).

Research Accuracy: Clinical trials can identify eligible patients through precise code clusters rather than chart reviews. A leukemia study seeking patients with specific genetic markers found 94% accurate identification using ICD-11 clusters versus 62% with ICD-10 codes.

Quality Measurement: Healthcare systems can track outcomes for precisely defined populations. Instead of measuring "diabetes control," they can examine "Type 2 diabetes with nephropathy, stage 3, on insulin therapy"—enabling meaningful quality improvement.

Case Example 2: Trauma Documentation

Emergency physician Dr. Michael Roberts treats a motorcycle accident victim. The patient has multiple injuries requiring comprehensive documentation for trauma registry, insurance, and follow-up care.

ICD-10 requires a frustrating list:

- S72.001A (Fracture of unspecified part of neck of right femur)
- S27.321A (Contusion of lung, unilateral)
- S06.0X0A (Concussion without loss of consciousness)
- V29.9XXA (Unspecified motorcycle accident)
- Multiple additional codes for each injury

ICD-11 tells the complete story efficiently:

```
Primary cluster:
```

```
NA23.4 (Fracture of neck of femur)&XK9J (Right)
/NB51.3 (Pulmonary contusion)&XK9J (Right)
/NA05.0 (Concussion)
&PC51.2E (Motorcycle driver injured in collision with
car)
&XU00.0 (Emergency department)
&XT11 (Initial encounter)
```

The cluster maintains relationships between injuries, mechanism, and encounter type—crucial for trauma research and prevention programs.

Digital-First Design Philosophy

ICD-11's digital foundation differs fundamentally from digitized paper systems. Every design decision assumes digital interaction, creating possibilities unimaginable in analog classification.

Real-Time Evolution

Traditional classifications freeze knowledge at publication. ICD-10, finalized in 1990, still classifies HIV as invariably fatal because that was true when written. Updates require cumbersome addenda that many systems never implement.

ICD-11 lives and breathes. The digital infrastructure supports:

- Monthly content updates for emerging diseases
- Annual structural revisions
- Real-time error corrections
- Immediate global distribution

When COVID-19 emerged, ICD-11 added specific codes within weeks. Variants received unique identifiers as they appeared. Contact tracing codes evolved with public health needs. The classification adapted as quickly as the pandemic itself (20).

API-Centric Architecture

Rather than distributing files, ICD-11 operates through APIs (Application Programming Interfaces). Think of APIs as universal translators enabling any system to speak ICD-11 fluently.

Healthcare organizations integrate ICD-11 APIs to:

- Search codes using natural language
- Validate combinations in milliseconds
- Retrieve definitions in preferred languages
- Access the latest version automatically

Dr. Jennifer Wu, Chief Medical Information Officer at a major health system, describes the transformation: "We used to spend months updating code tables. Now our EHR connects directly to WHO. Updates happen seamlessly. Our coders always work with current information" (21).

Case Example 3: Mental Health Innovation

A psychiatric clinic implementing ICD-11 discovered unexpected benefits from digital integration. Dr. Sarah Johnson explains their experience:

"We treat complex patients with multiple mental health conditions. ICD-10 forced us to prioritize—which diagnosis goes first? What gets left out? ICD-11's digital tools let us paint the complete picture."

For a patient with depression, anxiety, and gaming addiction:

```
6A70.3 (Recurrent depressive disorder, current episode
mild)
/6B73 (Generalized anxiety disorder)
/6C51 (Gaming disorder)
&Duration codes for each condition
&Severity scales updated at each visit
&Treatment response markers
```

The clinic's EHR pulls these codes through API, tracks changes over time, and generates outcome reports previously requiring manual chart review. Digital integration transforms routine documentation into research-quality data.

Intelligent Assistance

Digital-first design enables AI-powered coding assistance:

Natural Language Processing: Clinicians describe conditions in plain language; AI suggests appropriate codes Example: "Diabetes with eye problems" → 5A11/9B71.0Z (Type 2 diabetes with diabetic retinopathy)

Pattern Recognition: System learns from coding patterns, improving suggestions Example: After coding diabetes, system prompts for common complications

Error Prevention: AI identifies potential mistakes before submission Example: Coding pregnancy in elderly males triggers verification

Workflow Integration: Digital tools embed within clinical documentation Example: Problem lists automatically generate ICD-11 codes during note creation

Multilingual Capabilities

Language barriers fragment global health knowledge. A breakthrough treatment published in Mandarin might take years to influence English-speaking physicians. ICD-11 addresses this through revolutionary multilingual architecture.

True Semantic Translation

Previous ICDs translated English terms word-for-word, often creating confusion. Medical concepts don't always translate directly—what English calls "depression" encompasses different concepts in other languages.

ICD-11 separates concepts from language. Each disease entity exists as an abstract idea linked to culturally appropriate terms in each language. The Japanese term for depression reflects Japanese understanding while mapping to the same underlying concept as English "depression."

Dr. Carlos Mendoza, leading Spanish translation efforts, explains: "We're not translating words; we're ensuring concepts resonate correctly across cultures. Sometimes this means one English term becomes several Spanish options, each appropriate for different regions" (22).

Implementation Across Languages

Currently, ICD-11 operates fully in 14 languages:

- Arabic (with right-to-left interface support)
- Chinese (Simplified and Traditional)
- English, French, Spanish, Portuguese
- Russian, Turkish, Czech
- Additional languages under development

Each language receives equal treatment—no "primary" language with "translations." Updates occur simultaneously across all supported languages.

Cultural Competence in Classification

Multilingual capability extends beyond translation to cultural competence. Consider traditional medicine integration:

- Chinese medicine describes "wind-cold invasion"
- Ayurveda refers to "vata imbalance"
- Western medicine calls it "upper respiratory infection"

ICD-11's architecture links these concepts while respecting their distinct theoretical frameworks. Practitioners can document using familiar concepts while maintaining international compatibility.

Integration with Modern Health IT

ICD-11 anticipates a healthcare ecosystem where systems seamlessly exchange information. Its technical standards ensure compatibility with current and future health technologies.

FHIR Compatibility

Fast Healthcare Interoperability Resources (FHIR) represents healthcare's internet standard. ICD-11's native FHIR support enables:

- Direct embedding in clinical documents
- Automatic code validation
- Terminology service integration
- Cross-system interoperability

When a physician in Tokyo documents a diagnosis, a researcher in London can immediately include that case in a global study—with full semantic understanding despite language differences.

Electronic Health Record Evolution

Modern EHRs integrate ICD-11 throughout clinical workflows:

Problem List Management: Diagnoses automatically code to ICD-11 with post-coordination options appearing contextually

Clinical Decision Support: ICD-11 codes trigger relevant guidelines, order sets, and alerts

Quality Reporting: Precise coding enables accurate quality metrics without manual abstraction

Research Integration: Patients meeting study criteria based on ICD-11 codes receive automatic screening alerts

Blockchain and Distributed Health

Emerging technologies find ready partnership with ICD-11:

Blockchain Integration: Immutable diagnosis records using ICD-11 standards **Federated Learning**: AI models training across institutions using standardized ICD-11 data **Internet of Medical Things**: Devices reporting conditions using ICD-11 codes **Precision Medicine**: Genomic data linked to phenotypes through ICD-11 classification

Dr. Yuki Tanaka, researching rare diseases, describes the impact: "ICD-11's digital standards let us aggregate cases globally. A condition affecting one in a million people now has thousands of documented cases accessible for research. The digital architecture makes rare diseases visible" (23).

Embracing the Digital Future

ICD-11's features represent more than technical improvements—they embody healthcare's digital transformation. Post-coordination captures clinical complexity. Digital architecture enables continuous evolution. Multilingual capabilities unite global health knowledge. Modern IT integration ensures relevance for decades ahead.

As you implement these features, think beyond coding accuracy. Consider how precise documentation through post-coordination improves patient care. Appreciate how digital updates keep pace with medical advances. Recognize how multilingual support advances health equity. Understand how IT integration enables innovations we're only beginning to imagine.

The transition requires effort, but the destination justifies the journey. ICD-11 doesn't just classify disease—it creates a foundation for healthcare's digital future.

Key Takeaways

- Post-coordination allows building precise diagnostic descriptions by combining stem codes with extensions

- Cluster coding maintains relationships between related conditions in single expressions
- Digital-first design enables real-time updates versus static publications
- APIs provide direct system integration without manual file updates
- Intelligent assistance through AI improves coding accuracy and efficiency
- Multilingual architecture goes beyond translation to ensure cultural competence
- 14 languages currently supported with simultaneous updates across all versions
- FHIR compatibility ensures integration with modern health IT standards
- Built-in validation prevents illogical code combinations while enabling clinical flexibility

Chapter 5: Detailed Mapping Tables from ICD-10 to ICD-11

The journey from ICD-10 to ICD-11 resembles navigating between two cities using different street layouts. While both reach similar destinations, the routes differ significantly. Success requires understanding not just where codes moved, but why they relocated and how to handle complex transitions.

Understanding WHO Mapping Resources

The World Health Organization recognized that successful ICD-11 adoption hinges on accurate mapping from existing systems. They developed comprehensive resources, but using them effectively requires understanding their structure and limitations.

Types of Mapping Relationships

WHO mapping tables identify four distinct relationships between ICD-10 and ICD-11 codes:

One-to-One Mappings: Direct equivalents where concepts translate cleanly. Approximately 40% of codes fall into this category—seemingly straightforward but requiring careful verification.

One-to-Many Mappings: Single ICD-10 codes splitting into multiple ICD-11 options. These account for 35% of mappings and demand clinical judgment to select appropriately.

Many-to-One Mappings: Multiple ICD-10 codes consolidating into single ICD-11 codes. Roughly 15% follow this pattern, simplifying documentation but requiring attention to lost granularity.

Complex Mappings: Situations requiring post-coordination or having no direct equivalent. The remaining 10% need careful analysis and often clinical input.

Dr. Patricia Anderson, who led mapping validation efforts, observes: "People expect a simple lookup table. Reality is far more nuanced. Each mapping tells a story about evolving medical understanding" (24).

Case Example 1: Diabetes Mapping Complexity

Consider diabetes classification evolution. ICD-10's E11.9 (Type 2 diabetes mellitus without complications) seems simple enough. But ICD-11 transforms this single code into multiple possibilities:

- 5A11.0 (Type 2 diabetes mellitus, blood glucose not well controlled)
- 5A11.1 (Type 2 diabetes mellitus, blood glucose well controlled)
- 5A11.2 (Type 2 diabetes mellitus, blood glucose control undetermined)

The mapping forces clinical documentation improvement. Dr. James Chen explains: "ICD-10 let us code diabetes without mentioning control status. ICD-11 demands this clinically relevant detail. It's better medicine, but requires workflow changes" (25).

For diabetes with complications, complexity multiplies:

- ICD-10: E11.21 (Type 2 diabetes with diabetic nephropathy)
- ICD-11: 5A11 (Type 2 diabetes) post-coordinated with 5A24.0 (Diabetic kidney disease)

This shift from pre-coordinated to post-coordinated coding affects:

- Documentation requirements
- Billing system capabilities
- Quality reporting algorithms
- Clinical decision support rules

Accessing and Using WHO Tools

39

WHO provides multiple mapping resources:

ICD-11 Browser Mapping Tab: Each code displays ICD-10 equivalents with relationship types and usage notes. Real-time updates ensure currency.

Downloadable Mapping Tables: Excel files containing complete mappings for offline use. Updated quarterly with version tracking.

API Services: Programmatic access enabling automated mapping within applications. Returns structured data with confidence scores.

Mapping Assistant Tool: Interactive application guiding complex mapping decisions through clinical questions.

However, tools alone don't ensure success. Organizations must understand mapping principles and develop local governance for ambiguous situations.

Major Disease Category Mappings

Each ICD chapter presents unique mapping challenges reflecting specialty-specific classification evolution. Understanding these patterns helps predict and address implementation issues.

Infectious Diseases Evolution

Chapter 1 demonstrates how scientific advancement drives classification change. HIV/AIDS classification exemplifies this evolution:

ICD-10 used broad categories:

- B20 (HIV disease resulting in infectious diseases)
- B21 (HIV disease resulting in malignant neoplasms)
- B22 (HIV disease resulting in other specified diseases)

ICD-11 provides nuanced classification:

- 1C60.0 (HIV disease clinical stage 1)
- 1C60.1 (HIV disease clinical stage 2)
- 1C60.2 (HIV disease clinical stage 3)
- 1C60.3 (HIV disease clinical stage 4)

Plus post-coordination for manifestations, creating precise clinical pictures impossible with ICD-10.

Case Example 2: Tuberculosis Classification

Dr. Monica Patel, an infectious disease specialist, describes tuberculosis mapping challenges: "ICD-10 scattered TB codes across multiple categories. ICD-11 consolidates them logically, but migration requires understanding both systems deeply."

ICD-10 tuberculosis codes:

- A15-A19 (Tuberculosis infections)
- B90 (Sequelae of tuberculosis)
- Various organ-specific codes

ICD-11 reorganization:

- 1A10-1A14 (Active tuberculosis by site)
- 1A15-1A16 (Latent tuberculosis)
- 1A17 (Sequelae codes)
- Post-coordination for drug resistance

The consolidation improves clinical logic but requires reconfiguring reports, alerts, and quality measures built around ICD-10's structure.

Mental Health Transformation

Chapter 6 underwent revolutionary changes reflecting evolved understanding of mental health:

Depression Mapping:

- ICD-10: F32.0-F32.9 (Depressive episode by severity)
- ICD-11: 6A70.0-6A70.3 (Single episode depressive disorder with refined severity scales)

The mapping appears straightforward until examining details. ICD-11 adds:

- Psychotic features specification
- Anxious distress qualifiers
- Melancholic features
- Seasonal pattern markers

Each addition requires clinical documentation previously uncaptured or stored in free text.

Neurodevelopmental Conditions: Autism spectrum mapping reveals classification philosophy changes:

- ICD-10: F84.0 (Childhood autism), F84.1 (Atypical autism), F84.5 (Asperger syndrome)
- ICD-11: 6A02 (Autism spectrum disorder) with intellectual and language impairment qualifiers

This consolidation reflects clinical consensus that these represent one condition with varying presentations rather than distinct disorders (26).

Complex Mapping Scenarios

Some transitions defy simple mapping, requiring sophisticated approaches to maintain clinical accuracy and operational continuity.

Multi-Axial to Linear Transitions

ICD-10's multi-axial approach to certain conditions disappears in ICD-11, replaced by post-coordination:

Injury Coding Evolution: ICD-10 required:

- Injury code (S00-T88)
- External cause code (V00-Y99)
- Place of occurrence code (Y92)
- Activity code (Y93)

ICD-11 integrates through clustering:

- Single injury code
- Post-coordinated with unified external cause description
- Extension codes capturing all contextual details

Case Example 3: Complex Injury Mapping

An emergency department treating a construction worker who fell from scaffolding faces mapping complexity:

ICD-10 documentation:

- S72.001A (Fracture of unspecified part of neck of right femur, initial encounter)
- W13.9XXA (Fall from, out of or through building, not otherwise specified)
- Y92.61 (Building under construction as place of occurrence)
- Y93.H3 (Activity, building and construction)
- Y99.0 (Civilian activity done for income or pay)

ICD-11 consolidation:

```
NA25.3 (Fracture of neck of femur)
&XK9J (Right side)
&PC50.0Y (Fall from scaffolding at construction site)
&XE6AT (While working for income)
&XT11 (Initial encounter)
```

The mapping requires understanding both systems' logic while maintaining data quality for injury surveillance programs expecting ICD-10 format.

Pregnancy and Childbirth Complications

Obstetric coding presents particular challenges due to timing and condition relationships:

ICD-10's O codes include temporal specifications:

- O24.0 (Pre-existing type 1 diabetes in pregnancy)
- O24.1 (Pre-existing type 2 diabetes in pregnancy)
- O24.4 (Gestational diabetes)

ICD-11 separates underlying conditions from pregnancy status:

- 5A10 or 5A11 (Underlying diabetes type)
- Post-coordinated with JA00-JA04 (Pregnancy trimester)
- Extension codes for control status

This philosophical shift—coding the condition plus pregnancy rather than pregnancy-modified conditions—affects clinical documentation, billing logic, and quality reporting.

Using Mapping Tools Effectively

Successful mapping requires more than mechanical code conversion. Organizations must develop systematic approaches addressing clinical, operational, and technical dimensions.

Clinical Validation Processes

Pure automated mapping fails because medical context matters. Successful organizations implement clinical review:

Specialty-Specific Teams: Cardiologists review cardiac mappings, oncologists validate cancer codes. Their clinical expertise catches nuances automated systems miss.

Use Case Testing: Rather than reviewing codes in isolation, teams test mappings against actual patient scenarios from their practice.

Edge Case Documentation: Complex or ambiguous mappings get documented with decision rationale, creating organizational knowledge bases.

Dr. Robert Wilson, leading a large health system's transition, shares their approach: "We treated mapping as clinical quality improvement, not IT implementation. Every specialty owned their mappings. This took longer but prevented downstream problems" (27).

Technical Implementation Strategies

Dual Coding Periods: Organizations maintain both ICD-10 and ICD-11 codes during transition, enabling:

- Parallel run validation
- Historical data continuity
- Gradual workflow transition
- Safety net for uncertainties

Mapping Confidence Scores: Rather than binary mapped/not mapped status, sophisticated systems assign confidence levels:

- High confidence: Direct equivalents
- Medium confidence: Clinical review recommended
- Low confidence: Manual mapping required

Crosswalk Maintenance: Mappings evolve as understanding improves. Version control and update processes ensure consistency:

- Monthly reviews of low-confidence mappings
- Quarterly updates from WHO incorporated
- Annual comprehensive validation
- Documentation of local decisions

Quality Assurance Frameworks

Mapping quality directly impacts patient care, research validity, and financial accuracy. Robust QA processes include:

Statistical Validation: Comparing disease prevalence before and after mapping identifies potential systematic errors. Sudden drops in diabetes prevalence suggest mapping problems, not health improvements.

Clinical Reasonableness: Subject matter experts review mapped data for clinical sense. Finding zero pregnancy complications after mapping indicates problems requiring investigation.

Financial Impact Analysis: Revenue cycle teams analyze mapping effects on reimbursement, identifying codes requiring special attention.

Research Continuity: Researchers validate that longitudinal studies maintain integrity across the classification transition.

Navigating the Transition

Mapping from ICD-10 to ICD-11 requires more than technical code conversion—it demands understanding evolving medical knowledge, recognizing classification philosophy changes, and maintaining operational continuity. The WHO provides excellent tools, but organizational commitment to clinical validation, technical sophistication, and quality assurance determines success.

Remember that perfect mapping rarely exists. The goal isn't mechanical translation but capturing clinical intent accurately in the new system. This requires embracing ICD-11's capabilities while respecting ICD-10's operational legacy.

Key Takeaways

- Four mapping types exist: one-to-one (40%), one-to-many (35%), many-to-one (15%), and complex (10%)
- WHO provides multiple mapping resources including browser tools, downloadable tables, APIs, and interactive assistants

- Diabetes mapping demonstrates evolution from simple codes to granular classification requiring glucose control documentation
- Mental health mapping reflects conceptual changes like autism spectrum consolidation
- Complex scenarios like injuries require understanding philosophical shifts from multi-axial to post-coordinated approaches
- Clinical validation by specialty experts prevents automated mapping errors
- Dual coding periods enable safe transitions while maintaining operational continuity
- Quality assurance must encompass statistical, clinical, financial, and research perspectives
- Success requires treating mapping as clinical quality improvement, not just technical conversion

Chapter 6: Step-by-Step Transition Planning Guide

Transitioning to ICD-11 resembles renovating a hospital while maintaining full operations. Every department needs updates, yet patient care cannot pause. Success demands meticulous planning, staged implementation, and constant communication. Organizations that approach this as merely a coding update consistently fail—those recognizing it as organizational transformation succeed.

Pre-Implementation Assessment

Before charting your course, you must understand your starting position. Pre-implementation assessment reveals not just what needs changing, but what resources and constraints shape your journey.

Current State Analysis

Begin with comprehensive system inventory. Most organizations discover surprising complexity:

Clinical Documentation Systems: Beyond the obvious EHR, identify:

- Specialty-specific systems (radiology, pathology, behavioral health)
- Departmental databases maintaining separate diagnosis lists
- Research registries using ICD codes
- Quality reporting platforms
- Clinical decision support tools referencing diagnoses

Dr. Lisa Thompson, who led assessment at a 10-hospital system, recalls: "We found 47 different systems using ICD-10 codes. Some we didn't even know existed—a researcher's database from 2015, still actively collecting data. Missing any system would have broken critical workflows" (28).

Case Example 1: Hidden System Discovery

Regional Medical Center thought they had 12 systems using ICD-10. Systematic assessment revealed:

- Emergency department triage system with embedded codes
- Nursing education platform using diagnoses for case studies
- Pharmacy system triggering alerts based on ICD codes
- Employee health system for occupational medicine
- Contract research organization interfaces
- State reporting systems for infectious diseases
- Insurance pre-authorization platforms

Each system required different transition approaches. The pharmacy system needed immediate updates for medication safety alerts. The education platform could transition gradually. Missing any would have caused operational failures.

Workflow Documentation

Understanding code usage patterns prevents disruption:

Clinical Workflows:

- How physicians document diagnoses
- When coding occurs (concurrent vs. retrospective)
- Who assigns codes (physicians, coders, automated systems)
- Review and correction processes

Operational Workflows:

- Billing cycle integration
- Quality measure abstraction
- Report generation timing
- External data submissions

Technical Workflows:

- Data flow between systems
- Update frequencies
- Validation processes
- Error handling procedures

Map these workflows visually. Flowcharts reveal dependencies invisible in written procedures.

Stakeholder Analysis

ICD-11 touches everyone in healthcare. Systematic stakeholder identification ensures no group gets overlooked:

Direct Users:

- Physicians documenting diagnoses
- Professional coders assigning classifications
- Quality analysts generating reports
- Researchers querying databases
- Billing staff processing claims

Indirect Users:

- Nurses using coded problem lists
- Pharmacists receiving diagnosis-triggered alerts
- Case managers reviewing coded conditions
- Administrators analyzing coded data
- Patients viewing coded diagnoses in portals

External Stakeholders:

- Insurance companies receiving claims
- Government agencies collecting data
- Research collaborators sharing information
- Health information exchanges
- Accreditation organizations

Each stakeholder group requires tailored communication and training strategies.

Creating Your Implementation Roadmap

With assessment complete, create a roadmap balancing ambitious progress with operational reality. The most elegant plans fail when they ignore organizational capacity.

Governance Structure

Successful implementations establish clear governance early:

Executive Steering Committee

- C-suite sponsor (crucial for resource allocation)
- Medical staff leadership
- Nursing leadership
- IT leadership
- Finance leadership
- Quality/Safety leadership

This committee makes strategic decisions, resolves conflicts, and ensures organizational alignment.

Implementation Leadership Team

- Project manager (full-time for large organizations)
- Clinical informaticist
- Coding manager
- IT technical lead
- Training coordinator
- Change management specialist

This team handles day-to-day implementation activities.

Clinical Advisory Groups

- Specialty-specific physician champions
- Nursing representatives
- Allied health professionals
- Front-line user representatives

These groups ensure clinical relevance and user acceptance.

Case Example 2: Governance in Action

Metro Health System's governance structure proved its worth when conflicts arose. The emergency department wanted rapid implementation for trauma registry requirements. Oncology demanded deliberate transition to maintain research continuity.

The Executive Steering Committee negotiated a solution:

- Emergency department piloted post-coordination for trauma
- Oncology maintained dual coding for 18 months
- Lessons from ED pilot informed oncology transition
- Both departments achieved their critical goals

Without governance, this conflict might have derailed implementation.

Phased Implementation Strategy

Rather than "big bang" conversion, successful organizations phase implementation:

Phase 1: Foundation (Months 1-6)

- Establish governance
- Complete assessment
- Select vendor partners
- Develop training strategy
- Create communication plan
- Design technical architecture

Phase 2: Preparation (Months 7-12)

- Upgrade technical infrastructure
- Develop system interfaces
- Create mapping tables
- Design new workflows
- Pilot training programs
- Begin stakeholder communication

Phase 3: Limited Production (Months 13-18)

- Implement in selected departments
- Dual code for validation
- Refine workflows based on experience
- Expand training programs
- Monitor quality metrics
- Adjust based on lessons learned

Phase 4: Full Implementation (Months 19-24)

- Organization-wide deployment
- Transition from dual coding
- Optimize workflows
- Advanced feature adoption
- Performance monitoring
- Continuous improvement

This timeline adjusts based on organizational size and complexity. Small practices might compress to 12-18 months; large academic centers may need 36-48 months.

Risk Management Integration

Every implementation faces risks. Acknowledging and planning for them improves success probability:

Technical Risks:

- Vendor readiness delays
- Integration complexity underestimation

- Performance degradation
- Data quality issues

Mitigation: Vendor agreements with penalties, proof-of-concept testing, performance benchmarking, data quality dashboards

Clinical Risks:

- Physician resistance
- Documentation quality decline
- Patient safety concerns
- Workflow disruption

Mitigation: Physician champions, gradual rollout, safety monitoring, workflow optimization teams

Financial Risks:

- Budget overruns
- Productivity losses
- Revenue cycle disruption
- Denied claims increases

Mitigation: Contingency funding, productivity support, parallel processing capability, payer communication

Case Example 3: Risk Becomes Reality

University Hospital planned meticulously but encountered unexpected risk. Their largest commercial payer announced they wouldn't accept ICD-11 codes for two years beyond the hospital's go-live date.

Risk mitigation strategies activated:

- Crosswalk engine translated ICD-11 to ICD-10 for that payer
- Dual coding continued for those claims
- Executive negotiations with payer leadership
- Financial reserves covered additional coding costs

- Eventually, payer pressure from other providers accelerated their timeline

Without risk planning, this could have forced implementation delays or financial crisis.

Resource Planning and Budgeting

ICD-11 implementation requires significant investment. Organizations consistently underestimate needs, leading to stressed implementations and suboptimal outcomes. Realistic budgeting prevents these problems.

Human Resource Requirements

Calculate staffing needs across multiple dimensions:

Dedicated Project Team:

- Project Manager: 1.0 FTE for 24+ months
- Clinical Lead: 0.5-1.0 FTE for 18+ months
- Technical Lead: 1.0 FTE for 24+ months
- Training Coordinator: 1.0 FTE for 18+ months
- Analysts: 2-4 FTE for 12+ months

Department Champions:

- Physician champions: 0.1-0.2 FTE per major department
- Nursing champions: 0.1 FTE per major unit
- Coding champions: 0.2-0.5 FTE

Productivity Impact:

- Coder productivity: 20-30% reduction for 6 months
- Physician documentation: 10-15 minutes additional daily for 3 months
- Quality analyst time: 25% increase for 12 months

Dr. Michael Chen, CFO of a health system, advises: "We budgeted for productivity loss but underestimated duration. Plan for longer impact than seems reasonable—you'll need it" (29).

Technology Investments

Beyond software licenses, consider full technology costs:

Infrastructure:

- Server upgrades for increased processing
- Network bandwidth for API traffic
- Storage expansion for dual coding period
- Backup system enhancements

Software Costs:

- EHR upgrade licensing
- Interface engine modifications
- Reporting tool updates
- Training system licenses
- Testing environment creation

Integration Expenses:

- Vendor professional services
- Interface development
- Custom report creation
- Workflow automation tools

Hidden Technology Costs: Organizations often miss:

- Legacy system modifications
- Third-party tool updates
- Increased help desk support
- Performance optimization services
- Security assessment updates

Risk Management Strategies

Effective risk management transforms potential disasters into manageable challenges. The key lies in proactive identification and systematic mitigation.

Clinical Risk Management

Patient safety remains paramount during transition:

Documentation Quality Monitoring:

- Daily audits of coded diagnoses
- Alert systems for unusual patterns
- Physician feedback on coding accuracy
- Patient safety event tracking

Decision Support Adaptation:

- Parallel run clinical alerts
- Validation of medication interactions
- Quality measure continuity
- Order set functionality verification

Communication Protocols:

- Clear escalation pathways
- Daily huddles during go-live
- Physician hotline for questions
- Real-time issue resolution

Operational Risk Management

Revenue Cycle Protection:

- Maintain ICD-10 crosswalks
- Daily claim rejection monitoring
- Payer readiness verification

- Denial pattern analysis
- Cash reserve planning

Compliance Assurance:

- Regulatory requirement mapping
- Audit trail maintenance
- Documentation standards updating
- External audit preparation

Technical Risk Management

System Performance:

- Baseline performance metrics
- Capacity planning models
- Load testing protocols
- Rollback procedures
- Disaster recovery updates

Data Integrity:

- Mapping validation processes
- Historical data preservation
- Audit capabilities
- Reconciliation procedures

Building Your Success Foundation

Successful ICD-11 implementation requires treating the transition as organizational transformation, not technical upgrade. Thorough assessment reveals hidden complexities. Thoughtful governance ensures coordinated progress. Realistic resource planning prevents crisis. Proactive risk management maintains stability.

Start planning now, even if implementation seems distant. Early preparation enables smooth transition, while rushed implementations guarantee struggles. Your patients deserve the benefits ICD-11 offers—clinical precision, operational efficiency, and improved outcomes. Careful planning ensures they receive them without disruption to care.

Key Takeaways

- Comprehensive assessment reveals 3-4x more affected systems than initially identified
- Governance structure must include executive, operational, and clinical perspectives
- Phased implementation over 18-36 months reduces risk versus "big bang" approaches
- Human resource needs include dedicated project staff plus department champions
- Budget for 20-30% coder productivity reduction lasting 6+ months
- Technology costs extend beyond software to infrastructure and integration
- Risk management must address clinical, operational, financial, and technical dimensions
- Early payer engagement prevents revenue cycle surprises
- Success requires treating implementation as organizational transformation

Chapter 7: Implementation Timelines and Strategies

The path from decision to successful ICD-11 operation varies dramatically across organizations. A small physician practice might complete transition in twelve months, while an academic medical center requires four years. Understanding these variations—and learning from others' experiences—shapes realistic timelines and effective strategies.

Phase-Based Implementation Approach

Successful implementations follow predictable phases, though duration and intensity vary by organizational complexity. Think of these phases as seasons—each has characteristic activities and challenges.

Phase 1: Foundation (Months 1-6)

This phase establishes everything necessary for success. Organizations often rush through, eager to see tangible progress. Resist this temptation—weak foundations cause implementation collapse.

Essential Activities:

Organizational Readiness Create burning platform for change. Staff need compelling reasons to embrace disruption. Dr. Patricia Williams, who led three hospital transitions, emphasizes: "Show clinicians how ICD-11 improves patient care, not just coding accuracy. When physicians see better clinical documentation possibilities, resistance melts" (30).

Vendor Assessment Evaluate all technology partners:

- EHR vendor ICD-11 roadmap
- Coding software capabilities

- Interface engine requirements
- Reporting tool updates
- Training platform options

Many organizations discover vendor unreadiness. Early identification enables vendor changes or workaround planning.

Infrastructure Planning (continued)

- Network bandwidth for increased API traffic
- Database optimization requirements
- Backup and recovery enhancements

Case Example 1: Foundation Phase Success

Community Medical Group (45 physicians) succeeded through methodical foundation building. They discovered their billing company wasn't planning ICD-11 support. This early identification allowed them to:

- Negotiate transition requirements into existing contract
- Establish clear timelines with penalties
- Create contingency plans for billing company transition
- Avoid revenue cycle disruption at go-live

Their meticulous foundation work prevented a crisis that would have emerged during implementation.

Phase 2: Design and Build (Months 7-12)

With foundations solid, construction begins. This phase transforms plans into reality.

Workflow Redesign Clinical workflows require optimization for ICD-11's capabilities:

- Post-coordination integration into documentation
- Code selection at point of care

- Quality review processes
- Error correction procedures

Involve front-line users throughout design. Their insights prevent theoretical workflows that fail practically.

Technical Development IT teams build:

- System interfaces
- Mapping tables
- Validation rules
- Report modifications
- Training environments

Parallel development streams require careful coordination. Regular integration testing prevents last-minute surprises.

Training Program Creation Develop role-specific curricula:

- Physician modules focusing on clinical benefits
- Coder training emphasizing post-coordination
- Analyst education on new report capabilities
- Administrator overviews of operational impacts

Adult learning principles apply—hands-on practice trumps lectures.

Phase 3: Testing and Validation (Months 13-15)

Testing separates successful implementations from disasters. Comprehensive testing takes time but prevents patient care disruption.

Technical Testing Layers:

- Unit testing (individual components)
- Integration testing (connected systems)
- Performance testing (response times)
- Security testing (access controls)
- Disaster recovery testing

Clinical Validation: Test with real scenarios from your practice:

- Common diagnoses
- Complex multi-condition patients
- Rare diseases
- Department-specific cases

Dr. Robert Chen describes their approach: "We pulled 100 actual cases from each department. Physicians coded them in ICD-11, compared to historical ICD-10 coding. Discrepancies revealed workflow issues and training needs" (31).

Phase 4: Pilot Implementation (Months 16-18)

Pilots provide real-world testing without organization-wide risk. Select pilot sites carefully:

- Engaged leadership
- Moderate complexity
- Representative workflows
- Resilient operations

Case Example 2: Pilot Reveals Hidden Issues

Large Academic Center piloted in dermatology—seemingly straightforward specialty. The pilot revealed:

- Biopsy workflows required complete redesign
- Integration with pathology systems needed updates
- Research database connections broke with new codes
- Provider documentation time increased 40%

These discoveries led to:

- Workflow optimization reducing documentation time to baseline
- Pathology system updates before widespread rollout
- Research database modifications

- Revised training emphasizing efficiency techniques

Without the pilot, these issues would have affected all departments simultaneously.

Phase 5: Staged Rollout (Months 19-24)

Full implementation proceeds in waves, building momentum while managing risk:

Wave Planning:

- Wave 1: Early adopters and simpler departments
- Wave 2: Medical specialties
- Wave 3: Surgical specialties
- Wave 4: Complex/specialized areas

Each wave incorporates lessons from predecessors, improving with experience.

Support Structure:

- Command center during each go-live
- Super-users providing at-the-elbow support
- Real-time issue tracking
- Daily optimization based on feedback

Country-Specific Experiences and Lessons

Global ICD-11 implementations provide valuable insights. Each country's approach reflects their healthcare system structure and priorities.

Thailand: Rapid National Implementation

Thailand achieved nationwide implementation in 18 months through:

- Strong Ministry of Health leadership

- Centralized health IT infrastructure
- Social media training campaigns
- Mobile-first approach for rural areas

Dr. Siriporn Tanasri, leading implementation, shares key insights: "We leveraged existing WhatsApp groups for peer support. Physicians helped each other learn, creating organic knowledge networks. Technology barriers disappeared when we met people where they already communicated" (32).

Lessons for Others:

- Existing communication channels outperform new platforms
- Peer learning accelerates adoption
- Mobile optimization essential for distributed workforces
- Government mandate helps but requires support

Germany: Methodical Regional Approach

Germany's federated system led to regional implementations:

- Pilot regions tested approaches
- Successful strategies replicated
- Regional variations accommodated
- Quality incentives drove adoption

The systematic approach took longer—four years for full implementation—but achieved deep integration and high quality.

Kuwait: Integration with National Health Strategy

Kuwait linked ICD-11 to broader health reforms:

- New hospital information systems included ICD-11 natively
- Medical education incorporated ICD-11 training
- Research infrastructure built on ICD-11 foundation
- Quality programs designed around ICD-11 capabilities

This strategic integration avoided treating ICD-11 as isolated change, embedding it in healthcare transformation.

Case Example 3: Finland's Translation Excellence

Finland faced unique challenges with complex language requiring precise medical terminology. Their approach:

- Two-year translation project involving clinicians and linguists
- Pilot testing in Swedish-speaking regions
- Cultural adaptation for Sami populations
- Validation through clinical scenarios

The investment in translation quality paid dividends:

- 95% clinician satisfaction with terminology
- Minimal clarification requests
- Faster adoption due to intuitive terms
- Model for other countries' translation efforts

Dual Coding Period Management

The transition period where both ICD-10 and ICD-11 operate simultaneously challenges organizations operationally and financially. Successful management strategies minimize burden while ensuring quality.

Duration Determination

Dual coding duration depends on multiple factors:

- Payer readiness (often the limiting factor)
- Research continuity requirements
- Quality measure transitions
- Regulatory mandates

Most organizations plan 12-18 months, though some extend to 24 months for complex situations.

Operational Strategies

Automated Assistance: Implement tools that:

- Generate ICD-11 codes from ICD-10 selections
- Suggest ICD-10 codes from ICD-11 documentation
- Flag mapping uncertainties for review
- Track mapping quality metrics

Workflow Optimization:

- Primary coding in ICD-11
- Automated crosswalk to ICD-10
- Exception-based review
- Quality sampling validation

Resource Management: Rather than double coding staff, organizations succeed through:

- Automation tools reducing manual work
- Offshore partners handling routine mappings
- Senior coders focusing on complex cases
- Continuous process improvement

Dr. Jennifer Martinez, coding director at a 400-bed hospital, reports: "We expected needing double our coding staff. Through automation and workflow optimization, we managed with 30% increase. The key was investing in technology rather than just adding people" (33).

Quality Assurance During Dual Coding

Maintaining accuracy across two systems challenges even strong organizations:

Reconciliation Processes:

- Daily automated comparison
- Pattern analysis for systematic issues

- Clinical review of discrepancies
- Continuous mapping refinement

Performance Metrics: Track both systems:

- Coding accuracy rates
- Productivity measures
- Denial rates by system
- Query rates by system

Communication Strategies: Keep stakeholders informed:

- Regular updates on transition progress
- Transparency about challenges
- Celebration of milestones
- Clear timelines for ICD-10 sunset

Success Factors from Early Adopters

Analyzing successful implementations reveals consistent patterns:

Leadership Commitment

Beyond project sponsorship, successful organizations show:

- C-suite visibility and involvement
- Board-level progress reporting
- Resource commitment despite competing priorities
- Clear messaging about strategic importance

Clinical Engagement

Physician involvement differentiates success from struggle:

- Physician champions in every department
- Clinical benefits emphasized over administrative
- Workflow design by users, not IT
- Continuous feedback incorporation

Change Management Excellence

Technical implementation without cultural change fails:

- Comprehensive communication strategies
- Stakeholder-specific messaging
- Regular pulse surveys
- Celebration of successes
- Transparent challenge acknowledgment

Investment in Training

Successful organizations invest heavily:

- Multiple learning modalities
- Role-specific content
- Hands-on practice opportunities
- Ongoing refreshers
- Competency validation

Continuous Improvement Mindset

Rather than "project completion," leaders frame as ongoing journey:

- Regular optimization cycles
- User feedback integration
- Emerging feature adoption
- Performance improvement focus

Case Example 4: Multi-System Success Story

Regional Health Network (12 hospitals, 200 clinics) succeeded through integrated approach:

Year 1: Foundation and planning

- Established shared governance across facilities
- Standardized workflows before ICD-11

- Created network-wide training programs
- Negotiated joint vendor agreements

Year 2: Phased implementation

- Piloted at flagship hospital
- Rolled out to community hospitals
- Extended to ambulatory clinics
- Integrated specialty facilities

Year 3: Optimization and advancement

- Implemented advanced features
- Achieved quality improvements
- Reduced coding costs 15%
- Improved research capabilities

Their keys to success:

- Treated as network initiative, not facility-by-facility
- Shared resources and lessons
- Standardized approaches where possible
- Accommodated necessary variations

Implementation timelines and strategies vary, but successful patterns emerge. Plan phases thoughtfully, allowing adequate time for each. Learn from global experiences while adapting to local contexts. Manage dual coding periods through automation and optimization. Most importantly, focus on success factors proven across diverse settings.

Your timeline will be unique, but your journey follows paths others have traveled. Their lessons light your way forward.

Key Takeaways

- Implementation typically requires 18-36 months across five distinct phases

- Foundation phase (months 1-6) determines overall success— never rush
- Pilot implementations reveal hidden issues before organization-wide impact
- Country experiences show multiple successful approaches: rapid national (Thailand), methodical regional (Germany), strategic integration (Kuwait)
- Dual coding periods average 12-18 months, managed through automation not staff doubling
- Success factors include C-suite commitment, physician engagement, and continuous improvement mindset
- Wave-based rollouts build momentum while managing risk
- Training investment and change management differentiate success from struggle
- Network approaches outperform facility-by-facility implementations

Chapter 8: Practice Scenarios and Case Studies

Real-world application of ICD-11 transforms theoretical knowledge into practical expertise. Through detailed scenarios across specialties, we'll explore how ICD-11's features solve longstanding documentation challenges while revealing implementation nuances that training alone cannot convey.

Specialty-Specific Coding Examples

Each medical specialty brings unique documentation needs. ICD-11's flexibility accommodates these differences while maintaining standardization. Let's examine how various specialties leverage ICD-11's capabilities.

Cardiology: Precision in Complexity

Modern cardiology manages increasingly complex patients with multiple interrelated conditions. ICD-11's post-coordination capabilities capture these relationships precisely.

Case Study: Advanced Heart Failure Management

Mrs. Eleanor Chen, 72, presents to the cardiology clinic with worsening dyspnea. Her medical history includes:

- Ischemic cardiomyopathy following anterior MI five years ago
- Chronic atrial fibrillation on anticoagulation
- Stage 3 chronic kidney disease
- Type 2 diabetes with good control
- Recent hospitalization for acute decompensated heart failure

ICD-10 coding required multiple unconnected codes:

- I50.23 (Acute on chronic systolic heart failure)

- I25.5 (Ischemic cardiomyopathy)
- I48.91 (Unspecified atrial fibrillation)
- N18.3 (Chronic kidney disease, stage 3)
- E11.9 (Type 2 diabetes without complications)
- Z87.74 (Personal history of sudden cardiac arrest)

ICD-11 tells an integrated story:

```
Primary: BD11.2 (Chronic heart failure with reduced
ejection fraction)
Post-coordinated with:
- BA41.Z (Previous myocardial infarction) / causing
condition
- BC81.4 (Persistent atrial fibrillation)
- Stage: Severity scale 3 (NYHA Class III)
- Temporal: Recent exacerbation

Secondary integrated conditions:
- GB61.3 (Chronic kidney disease, stage 3) / related to
cardiac condition
- 5A11.1 (Type 2 diabetes, well controlled)
```

Dr. Michael Rodriguez, interventional cardiologist, explains the clinical impact: "ICD-11 captures the relationship between her ischemic event and current heart failure. This connection affects treatment decisions—we know her cardiomyopathy is ischemic, not idiopathic. The staging information guides therapy intensity. Previously, these connections lived in narrative notes, invisible to quality metrics and research databases" (34).

Oncology: Tumor Board Precision

Cancer care increasingly relies on molecular characterization. ICD-11 accommodates genetic markers and targeted therapy considerations.

Case Study: Precision Oncology Documentation

James Patterson, 58, diagnosed with lung adenocarcinoma during evaluation of persistent cough. Comprehensive testing reveals:

- Right upper lobe primary tumor, 4.2 cm
- Ipsilateral hilar lymph node involvement
- EGFR mutation (exon 19 deletion)
- PD-L1 expression 60%
- No distant metastases

ICD-10's limitations frustrated oncologists:

- C34.11 (Malignant neoplasm of upper lobe, right bronchus or lung)
- C77.1 (Secondary malignant neoplasm of intrathoracic lymph nodes)
- No codes for molecular markers
- Z51.11 (Encounter for antineoplastic chemotherapy)

ICD-11 enables precision documentation:

```
2C25.1 (Adenocarcinoma of bronchus or lung)
Post-coordinated with:
- XK9J (Laterality: right)
- XA8PH7 (Upper lobe)
- Stage: T2N1M0 (Stage IIB)
- Molecular markers:
  - EGFR mutation positive (specific deletion)
  - PD-L1 high expression
- 2D11.1 (Regional lymph node metastasis)
```

This precise coding enables:

- Automatic clinical trial matching
- Targeted therapy selection algorithms
- Accurate prognosis calculation
- Quality metric stratification by molecular subtype

Dr. Sarah Kim, thoracic oncologist, notes: "ICD-11 finally speaks our language. We discuss patients by molecular subtypes, not just anatomic locations. The coding system now reflects how we actually practice precision oncology" (35).

Rheumatology: Capturing Disease Activity

Rheumatic diseases fluctuate, making point-in-time coding challenging. ICD-11's temporal qualifiers and severity scales address this need.

Case Study: Rheumatoid Arthritis Complexity

Dr. Patel treats Maria Santos, 45, with seropositive rheumatoid arthritis. Today's visit reveals:

- Active synovitis in MCPs and wrists bilaterally
- Morning stiffness lasting 90 minutes
- Elevated inflammatory markers (CRP 45, ESR 68)
- Subcutaneous nodules on elbows
- Secondary Sjögren's syndrome
- Medication: methotrexate and adalimumab

ICD-10 coding felt inadequate:

- M05.79 (Rheumatoid arthritis with rheumatoid factor without organ involvement)
- M35.00 (Sjögren's syndrome, unspecified)
- M06.30 (Rheumatoid nodule, unspecified site)

ICD-11 captures disease reality:

```
FA20.1 (Seropositive rheumatoid arthritis)
Post-coordinated with:
- Activity level: High disease activity (DAS28 > 5.1)
- Manifestations:
  - Joint involvement pattern (MCPs, wrists)
  - Extra-articular: Subcutaneous nodules
- Associated: 4A60.Z (Sjögren syndrome)
- Treatment status: On combination DMARD therapy
```

This detailed coding supports:

- Insurance approval for biological therapy

- Quality measures for treat-to-target strategies
- Research cohort identification
- Outcome tracking over time

Emergency Department Scenarios

Emergency departments face unique challenges—high acuity, time pressure, and incomplete information. ICD-11 must function efficiently in this environment.

Case Study: Multi-Trauma Documentation

Saturday night, trauma bay receives Thomas Wilson, 34, motorcycle versus automobile collision. The trauma team documents:

- Helmet worn, moderate damage
- Loss of consciousness at scene, now GCS 14
- Obvious right femur deformity
- Abdominal tenderness
- Multiple abrasions and lacerations

Initial imaging reveals:

- Right femur shaft fracture
- Small subdural hematoma
- Grade 2 splenic laceration
- Multiple rib fractures (right 4-7)

ICD-10 required extensive coding:

- S72.301A (Unspecified fracture of shaft of right femur, initial encounter)
- S06.5X1A (Traumatic subdural hemorrhage with loss of consciousness of 30 minutes or less)
- S36.031A (Moderate laceration of spleen)
- S22.42XA (Multiple fractures of ribs, left side)
- V23.4XXA (Motorcycle driver injured in collision with car)
- Multiple additional codes for circumstances

ICD-11 streamlines while improving precision:

```
Primary injury cluster:
NA25.2 (Fracture of shaft of femur) & XK9J (Right)
/ NA05.4 (Subdural haematoma) & XP45.2 (Brief LOC)
/ ND94.4 (Injury of spleen) & Grade 2
/ NA34.Z (Multiple rib fractures) & XK9J (Right, ribs 4-
7)

Causation cluster:
PC51.2D (Motorcycle rider collision with car)
& XP67.1 (Helmet worn)
& Speed and impact qualifiers
```

Dr. Jennifer Chang, trauma surgeon, appreciates the efficiency: "ICD-11's clustering maintains injury relationships. For trauma registry and quality improvement, understanding injury patterns matters more than isolated injuries. The external cause integration eliminates redundant documentation" (36).

Case Study: Diagnostic Uncertainty

Emma Thompson, 67, presents with chest pain and dyspnea. Initial evaluation suggests possible:

- Acute coronary syndrome
- Pulmonary embolism
- Aortic dissection

After extensive workup:

- Cardiac enzymes negative
- CT angiography negative for PE and dissection
- Diagnosis: Severe GERD with esophageal spasm

ICD-10 struggled with evolving diagnoses:

- Initial: R07.9 (Chest pain, unspecified)
- Final: K21.9 (Gastro-esophageal reflux disease without esophagitis)

77

- No capture of diagnostic journey

ICD-11 documents the complete encounter:

```
MG30.0Y (Chest pain)
Post-coordinated with:
- Differential diagnosis considered
- Ruled out: BA41 (Acute MI), BD40 (PE), BD50 (Aortic
dissection)
Final diagnosis: DA42.1 (GERD with esophageal symptoms)
& Temporal: Acute presentation
```

This coding supports:

- Quality metrics for diagnostic accuracy
- Appropriate testing documentation
- Risk stratification protocols
- Learning health system improvements

Complex Case Management

Patients with multiple chronic conditions challenge traditional coding systems. ICD-11's relationship modeling excels here.

Case Study: Geriatric Complexity

Dr. Williams manages Dorothy Johnson, 82, in her primary care medical home. Her conditions include:

- Moderate Alzheimer's dementia
- Frequent falls with previous hip fracture
- Chronic pain from lumbar stenosis
- Depression following husband's death
- Mild malnutrition
- Polypharmacy (12 medications)
- Lives alone, daughter provides support

ICD-10's problem list appears disconnected:

- F03.91 (Alzheimer's disease with behavioral disturbance)
- R29.6 (Repeated falls)
- M48.062 (Spinal stenosis, lumbar region with neurogenic claudication)
- F32.1 (Major depressive disorder, single episode, moderate)
- E46 (Unspecified protein-calorie malnutrition)

ICD-11 reveals interconnections:

```
Primary: 8A20.1 (Alzheimer disease, moderate severity)
Consequences and associations:
- PL20 (Repeated falls) / related to cognitive
impairment
- ME84.1 (Chronic primary pain) / affecting function
- 6A70.1 (Single episode depressive disorder) / related
to bereavement
- 6C80.2 (Mild protein-energy malnutrition) / related to
self-care deficit
- Functional status qualifiers
- Social determinant codes for isolation risk
```

Care coordinator Lisa Anderson describes the impact: "ICD-11 shows why Mrs. Johnson falls—it's not just aging, but cognitive impairment affecting judgment. Her malnutrition links to depression and functional decline. This coding guides our interventions toward root causes, not just symptoms" (37).

Mental Health Coding Improvements

Mental health classification underwent revolutionary changes in ICD-11, better reflecting clinical understanding and reducing stigma.

Case Study: Integrated Mental Health

Alex Rivera, 28, seeks help for multiple concerns:

- Mood swings affecting work performance
- Excessive gaming (14+ hours on weekends)
- Social anxiety limiting relationships
- Past trauma from childhood abuse

- Cannabis use "for anxiety"
- Questions about gender identity

ICD-10's categorical approach felt restrictive:

- F31.81 (Bipolar II disorder)
- F41.1 (Generalized anxiety disorder)
- F12.20 (Cannabis use disorder, uncomplicated)
- Z91.411 (Personal history of adult psychological abuse)

ICD-11 provides nuanced documentation:

```
6A71.1 (Bipolar disorder type II, current episode
depressive)
Co-occurring conditions:
- 6C51.Z (Gaming disorder) & severity moderate
- 6B23 (Social anxiety disorder)
- 6E50.Z (Cannabis use disorder, mild)
- QE60.7 (Personal history of psychological abuse in
childhood)

Associated factors:
- HA60.Z (Gender incongruence) / exploration phase
- Functional impairment: occupational and social
- Protective factors: employed, seeking treatment
```

Dr. Rachel Martinez, psychiatrist, explains the clinical relevance: "ICD-11 captures Alex's reality—multiple interrelated conditions, not hierarchical diagnoses. Gaming disorder recognition enables evidence-based treatment. Gender incongruence coding without mental disorder implications reduces stigma while supporting appropriate care" (38).

Case Study: Pediatric Neurodevelopment

Eight-year-old Ethan Chen's teacher raises concerns about classroom behavior. Comprehensive evaluation reveals:

- Difficulty sustaining attention
- Hyperactive behaviors in structured settings
- Reading significantly below grade level

- Excellent mathematics skills
- Social communication challenges
- Sensory sensitivities (noise, textures)

ICD-10 forced categorical decisions:

- F90.2 (Combined type ADHD)
- F81.0 (Specific reading disorder)
- Consider autism spectrum?

ICD-11 accommodates complexity:

```
6A05.Z (Attention deficit hyperactivity disorder)
& Presentation: Combined type
& Severity: Moderate impairment

Co-occurring:
- 6A03.1 (Developmental learning disorder) & specific to
reading
- 6A02.Z (Autism spectrum disorder) & without
intellectual impairment
- Sensory processing differences
- Strengths: mathematical ability above age level
```

This comprehensive coding:

- Guides educational accommodations
- Supports therapeutic interventions
- Avoids forced choosing between diagnoses
- Recognizes strengths alongside challenges

These scenarios demonstrate ICD-11's power to capture clinical reality. Post-coordination enables precision without overwhelming complexity. Relationship modeling reveals connections traditional coding obscures. Temporal and severity qualifiers track conditions over time. Most importantly, ICD-11 speaks clinicians' language, documenting how medicine is actually practiced.

Success requires shifting from "finding the right code" to "building the right description." This mindset change, supported by real-world practice, transforms ICD-11 from classification system to clinical communication tool.

Key Takeaways

- Post-coordination enables precise documentation of complex clinical scenarios
- Relationship modeling reveals connections between conditions affecting treatment
- Emergency departments benefit from injury clustering and integrated external causes
- Complex chronic disease management improves through documented interconnections
- Mental health coding reduces stigma while improving clinical accuracy
- Pediatric cases can capture multiple conditions without forced hierarchies
- Temporal and severity qualifiers track disease evolution over time
- Success requires shifting from code selection to description building
- ICD-11 documents medicine as actually practiced, not theoretical simplifications

Chapter 9: Coding Exercises with Answer Keys

Learning ICD-11 requires practice. Like mastering a musical instrument, theoretical knowledge must combine with hands-on experience. These exercises progress from basic skills to complex scenarios, building competence systematically. Work through each section before checking answers—struggle produces learning.

Basic Coding Practice

Start with straightforward cases to understand ICD-11's structure and basic post-coordination principles.

Exercise Set 1: Single Condition Coding

Code the following diagnoses in ICD-11:

1. Essential hypertension, well-controlled
2. Type 1 diabetes mellitus without complications
3. Acute appendicitis
4. Community-acquired pneumonia
5. Major depressive disorder, single episode, mild

Work through these before proceeding to answers.

Answer Key Set 1:

1. **Essential hypertension, well-controlled**
   o Code: BA00.Z
   o Note: ICD-11 doesn't require control status in the base code, but you can add extension codes for blood pressure readings or control status if needed
2. **Type 1 diabetes mellitus without complications**
   o Code: 5A10.Z

- o Note: The .Z indicates "unspecified" within the category. More specific codes exist if details available (e.g., 5A10.0 for poorly controlled)
3. **Acute appendicitis**
   - o Code: DB36.Z
   - o Further specificity available: DB36.1 (with localized peritonitis), DB36.2 (with generalized peritonitis), DB36.3 (with localized abscess)
4. **Community-acquired pneumonia**
   - o Code: CA40.0
   - o Note: Distinguish from hospital-acquired (CA40.1) or ventilator-associated (CA40.2)
5. **Major depressive disorder, single episode, mild**
   - o Code: 6A70.0
   - o The fourth character indicates severity: .0 (mild), .1 (moderate), .2 (severe without psychotic symptoms)

Exercise Set 2: Adding Basic Extensions

Now code these conditions with appropriate extensions:

1. Fracture of right ankle
2. Type 2 diabetes with poor control
3. Migraine headache, occurring monthly
4. Acute myocardial infarction, anterior wall
5. Mild cognitive impairment due to Alzheimer disease

Answer Key Set 2:

1. **Fracture of right ankle**
2. NC72.3 (Fracture of ankle)
3. & XK9J (Laterality: right side)
4. **Type 2 diabetes with poor control**
5. 5A11.0 (Type 2 diabetes mellitus, blood glucose not well controlled)

Or alternatively:

5A11.Z

```
& Blood glucose control status extension code
```

6. Migraine headache, occurring monthly
```
7. 8A80.Z (Migraine)
8. & Frequency qualifier: monthly episodes
```
9. Acute myocardial infarction, anterior wall
```
10.  BA41.0 (ST elevation myocardial infarction)
11.  & XA3KP4 (Anterior wall location)
```
12. Mild cognitive impairment due to Alzheimer disease
```
13.  8A20.0 (Alzheimer disease, mild severity)
14.  / 6D71.Z (Mild neurocognitive disorder)
```

Note: The forward slash (/) links the underlying disease with its manifestation

Post-coordination Exercises

These exercises build skill in combining codes to capture complex clinical scenarios.

Exercise Set 3: Disease Complications

Create complete post-coordinated codes for:

1. Type 2 diabetes with diabetic retinopathy and nephropathy
2. Rheumatoid arthritis with lung involvement
3. Cirrhosis due to chronic hepatitis C
4. Breast cancer with bone metastases
5. COPD with acute exacerbation

Answer Key Set 3:

1. Type 2 diabetes with diabetic retinopathy and nephropathy
```
2. 5A11.Z (Type 2 diabetes mellitus)
3. / 9B71.0Z (Diabetic retinopathy)
4. / 5A25.Z (Diabetic kidney disease)
```

Additional specificity possible with staging for each complication

5. **Rheumatoid arthritis with lung involvement**
6. `FA20.1 (Seropositive rheumatoid arthritis)`
7. `/ CB02.Z (Interstitial lung disease)`
8. `& Causation: secondary to RA`
9. **Cirrhosis due to chronic hepatitis C**
10. `1E51.1 (Chronic hepatitis C)`
11. `/ DB94.Z (Cirrhosis of liver)`

The primary condition (hepatitis C) comes first, followed by its consequence

12. **Breast cancer with bone metastases**
13. `2C61.Z (Malignant neoplasm of breast)`
14. `/ 2D12.3 (Secondary malignant neoplasm of bone)`
15. `& Site specification if known (e.g., vertebrae, femur)`
16. **COPD with acute exacerbation**
17. `CA22.0 (Chronic obstructive pulmonary disease with acute exacerbation)`

Note: This common scenario has a pre-coordinated code in ICD-11

Exercise Set 4: Complex Clustering

Build complete code clusters for these scenarios:

1. Motor vehicle accident resulting in:
   o Concussion with 2-hour loss of consciousness
   o Fractured ribs (left 3-5)
   o Splenic laceration
   o Driver was texting
2. Elderly patient with:
   o Moderate Alzheimer dementia
   o Recurrent falls
   o Dehydration
   o Living alone
3. Chemotherapy encounter for:
   o Stage IIIB lung adenocarcinoma
   o EGFR mutation positive

    o   Previous radiation therapy

Answer Key Set 4:

1. **Motor vehicle accident trauma cluster**
2. Primary cluster:
3. NA05.0 (Concussion)
4. & XP45.3 (LOC 1-6 hours)
5. / NA34.Z (Multiple rib fractures)
6. & XK8H (Left side, ribs 3-5)
7. / ND94.4 (Injury of spleen)
8. & Severity qualifier
9.
10. External cause:
11. PC50.1A (Car driver injured in collision)
12. & XP90.2 (Distracted by electronic device)
13. & XT11 (Initial encounter)
14. **Geriatric syndrome cluster**
15. 8A20.1 (Alzheimer disease, moderate severity)
16. Consequences:
17. / PL20 (Repeated falls)
18. & Related to cognitive impairment
19. / 6C70.Z (Dehydration)
20. & Risk factor: self-care deficit
21.
22. Social determinant:
23. QD70.2 (Lives alone)
24. & Support need identified
25. **Oncology treatment encounter**
26. Primary reason:
27. QA10.Z (Radiotherapy session)
28.
29. For condition:
30. 2C25.1 (Adenocarcinoma of lung)
31. & Stage IIIB (T4N2M0)
32. & EGFR mutation positive
33. & XH70.1 (Previous chemotherapy)

Specialty-Specific Challenges

These exercises address nuances within specific medical specialties.

Exercise Set 5: Cardiology Focus

Code these cardiac scenarios:

1. Acute decompensated heart failure due to ischemic cardiomyopathy, EF 25%
2. Atrial fibrillation with rapid ventricular response, newly diagnosed
3. NSTEMI with multivessel disease, pending CABG

Answer Key Set 5:

1. **Acute decompensated heart failure**
2. BD11.1 (Acute heart failure)
3. Due to:
4. BD91.Z (Ischemic cardiomyopathy)
5. & Ejection fraction 25%
6. & NYHA Class III-IV symptoms
7. **New atrial fibrillation with RVR**
8. BC81.0 (Paroxysmal atrial fibrillation)
9. & Rapid ventricular response
10. & New onset
11. & Consider adding stroke risk scores
12. **NSTEMI with multivessel disease**
13. BA42.Z (Non-ST elevation myocardial infarction)
14. & Multivessel coronary disease
15. & Planned intervention: CABG
16. & Risk stratification scores if documented

Common Error Correction

Learning from mistakes accelerates mastery. These exercises highlight frequent coding errors.

Exercise Set 6: Identify and Correct Errors

Find and fix the errors in these coding attempts:

1. **Incorrect**: 5A11.Z & XK9J **Scenario**: Type 2 diabetes **Error**: Laterality extension inappropriate for diabetes
2. **Incorrect**: BA00.Z / BA41.Z **Scenario**: Hypertension and separate MI **Error**: Forward slash implies causal relationship

3. **Incorrect**: 6A70.5 **Scenario**: Severe depression **Error**: Invalid fourth character
4. **Incorrect**: NA25.2 (Femur fracture) & COPD **Scenario**: Patient has both conditions **Error**: Ampersand links extensions to stem, not separate conditions
5. **Incorrect**: 2C61.Z & Male patient **Scenario**: Breast cancer in male **Error**: No error—males can have breast cancer

Answer Key Set 6:

1. **Corrected**: 5A11.Z (standalone, no laterality needed)
2. **Corrected**:
3. BA00.Z (Essential hypertension)
4. Separate code: BA41.Z (Acute MI)

   Use forward slash only for causal relationships

5. **Corrected**: 6A70.2 (Severe depression uses .2, not .5)
6. **Corrected**:
7. NA25.2 (Femur fracture) with appropriate extensions
8. Separate code: CA22.Z (COPD)
9. **No correction needed**—coding is appropriate

Case Challenge: Complete Documentation

Apply all skills to this complex case:

Patient: Sarah Martinez, 54-year-old teacher **Presenting complaint**: Worsening fatigue and joint pain **History**:

- Lupus diagnosed 5 years ago
- Recent kidney biopsy showing Class IV lupus nephritis
- Hypertension (on treatment)
- Depression following lupus diagnosis
- Raynaud's phenomenon
- Current medications: Prednisone, hydroxychloroquine, mycophenolate

Today's findings:

- Active joint inflammation (wrists, MCPs)
- Malar rash present
- BP 145/92 on medications
- Creatinine elevated (1.8)
- Discussing cyclophosphamide therapy

Create a complete ICD-11 code cluster for this encounter.

Case Challenge Answer:

```
Primary condition:
4A60.1 (Systemic lupus erythematosus)
& Disease activity: High (SLEDAI > 12)

Organ involvement:
/ GB45.Z (Lupus nephritis)
& Class IV (diffuse proliferative)
& Creatinine 1.8

Active manifestations:
/ ME55.Z (Arthritis)
& Sites: wrists, MCPs bilateral
/ EH10.Z (Malar rash)
/ BD11.Z (Raynaud phenomenon)

Comorbidities:
BA00.Z (Essential hypertension)
& On treatment, suboptimal control

6A70.1 (Depressive disorder)
& Related to chronic disease adjustment

Treatment consideration:
Under evaluation for cyclophosphamide
```

These exercises demonstrate ICD-11's flexibility and precision. Initial attempts may feel awkward—this is normal. With practice, post-coordination becomes intuitive. Focus on capturing clinical intent rather than perfect syntax initially. Accuracy improves with experience.

Continue practicing with cases from your specialty. Build personal libraries of common scenarios. Share challenging cases with colleagues. Most importantly, remember that ICD-11 coding is a tool for better clinical documentation, not an end in itself.

Key Takeaways

- Start with single condition coding before attempting post-coordination
- Forward slash (/) indicates causal relationships or manifestations
- Ampersand (&) adds extensions to describe the stem code
- Common errors include inappropriate extensions and incorrect relationship indicators
- Specialty-specific patterns emerge with practice
- Complex cases require systematic approaches—primary condition first
- Build code clusters that tell complete clinical stories
- Error analysis accelerates learning
- Regular practice with real cases builds expertise faster than theoretical study

Chapter 10: Common Pitfalls and How to Avoid Them

Experience teaches harsh lessons. Organizations worldwide have discovered ICD-11 implementation pitfalls through painful trial and error. By studying their struggles, you can navigate around these obstacles rather than stumbling into them. Prevention costs far less than correction.

Top 10 Implementation Mistakes

1. Treating ICD-11 as a Simple Upgrade

The most dangerous misconception views ICD-11 as merely "ICD-10 with new codes." This fundamental misunderstanding cascades into every implementation decision.

The Reality: ICD-11 represents a paradigm shift in health information management. Post-coordination changes documentation workflows. Digital architecture requires infrastructure updates. Semantic relationships affect reporting logic.

Dr. Thomas Anderson, who salvaged a failing implementation, recalls: "Our CEO called it a 'coding update' and delegated to IT. Six months later, physicians revolted, claims were denied, and quality reports failed. We had to stop and restart, treating it as the transformation it truly is" (39).

Prevention Strategy:

- Frame as organizational transformation from day one
- Engage clinical leadership immediately
- Budget for comprehensive change, not technical updates
- Communicate the fundamental differences repeatedly

Case Example: Course Correction

Metropolitan Health initially allocated three months and $200,000 for ICD-11 "upgrade." After near-disaster, they:

- Extended timeline to 24 months
- Increased budget to $2.8 million
- Appointed physician executive as sponsor
- Treated as strategic initiative

The restart succeeded where the initial attempt failed catastrophically.

2. Underestimating Timeline Requirements

Optimism bias affects most implementation timelines. Organizations see vendor promises and best-case scenarios, ignoring typical experiences.

Common Timeline Mistakes:

- Planning 6-12 months for major medical centers
- Assuming vendor readiness equals organizational readiness
- Ignoring dual coding period requirements
- Underestimating training duration

Realistic Timelines by Organization Type:

- Solo practice: 6-9 months
- Small group (5-20 physicians): 9-12 months
- Community hospital: 18-24 months
- Multi-hospital system: 24-36 months
- Academic medical center: 36-48 months

Prevention Strategy:

- Study similar organizations' experiences
- Add 50% buffer to initial estimates
- Plan phases with go/no-go decision points
- Build in time for optimization post-implementation

3. Inadequate Stakeholder Engagement

IT-led implementations consistently struggle. Success requires active engagement across all affected groups.

Commonly Overlooked Stakeholders:

- Night shift nurses using coded problem lists
- Researchers with longitudinal studies
- Quality analysts generating reports
- External partners receiving data feeds
- Patients viewing codes in portals

Case Example: Research Disaster Avoided

University Medical Center discovered, two weeks before go-live, that ICD-11 would break 47 active research protocols. A junior analyst's question during training revealed the issue. Emergency stakeholder engagement found:

- $12 million in research grants at risk
- 3,000 enrolled patients affected
- FDA submissions in jeopardy

Rapid response included:

- Crosswalk development for research databases
- Dual coding for research patients
- Protocol amendments filed
- Disaster avoided through late stakeholder discovery

Prevention Strategy:

- Map all code touchpoints systematically
- Include representatives in planning
- Conduct impact assessments by stakeholder group
- Maintain ongoing communication channels

4. Insufficient Training Investment

Organizations consistently underestimate training needs, focusing on technical skills while ignoring workflow changes and clinical concepts.

Training Failure Patterns:

- One-time training sessions
- Generic content for all roles
- Focus on code lookup mechanics
- No competency validation
- Absent ongoing education

Dr. Patricia Wong, medical informaticist, explains: "We trained people to use the tools but not think in ICD-11. Coders memorized post-coordination syntax but couldn't identify when to use it. Physicians learned code searches but not documentation changes. Technical training without conceptual understanding fails" (40).

Comprehensive Training Framework:

Phase 1: Conceptual Foundation (Before system access)

- Why ICD-11 differs fundamentally
- Post-coordination principles
- Clinical documentation improvements
- Role-specific benefits

Phase 2: Technical Skills (With system access)

- Navigation and search techniques
- Post-coordination mechanics
- Workflow integration
- Error recognition

Phase 3: Applied Practice (Pre-implementation)

- Department-specific scenarios
- Common challenge areas
- Peer learning sessions
- Competency validation

Phase 4: Ongoing Development (Post-implementation)

- Monthly refreshers
- New feature training
- Optimization techniques
- Advanced capabilities

5. Ignoring Workflow Redesign

ICD-11's capabilities require workflow evolution. Organizations forcing new codes into old workflows sacrifice major benefits.

Workflow Assumptions That Fail:

- Physicians will document the same way
- Coding happens at the same points
- Review processes remain unchanged
- Reporting logic stays constant

Case Example: Emergency Department Revolution

Regional Trauma Center initially maintained traditional workflow:

1. Physician documents in free text
2. Coder interprets documentation later
3. Codes assigned retrospectively
4. Queries for missing information

This approach collapsed with ICD-11's specificity requirements. Redesigned workflow:

1. Structured documentation captures ICD-11 elements
2. Real-time coding assistance during documentation

3. Automated post-coordination suggestions
4. Exception-based review only

Results:

- Documentation time decreased 15%
- Coding accuracy improved 30%
- Query rate dropped 60%
- Physician satisfaction increased

6. Poor Communication Strategies

Change resistance multiplies with poor communication. Organizations often communicate too little, too late, through wrong channels.

Communication Failures:

- Technical jargon alienating clinicians
- One-size-fits-all messaging
- Emphasis on compliance versus benefits
- Lack of two-way communication
- Insufficient frequency

Effective Communication Framework:

Audience-Specific Messaging:

- Physicians: Clinical documentation improvements
- Nurses: Care coordination benefits
- Administrators: Operational efficiency gains
- Patients: More accurate health records

Multi-Channel Approach:

- Leadership town halls
- Department meetings
- Email updates (brief, visual)
- Intranet resources

- Peer champions
- Quick reference cards

Timing Strategy:

- 18 months out: Awareness building
- 12 months out: Detailed planning communication
- 6 months out: Training announcements
- 3 months out: Countdown communications
- Go-live: Daily updates
- Post-live: Success stories and optimization

7. Vendor Dependency Without Verification

Trusting vendor readiness without verification creates implementation crises.

Common Vendor Issues:

- Promised features not delivered
- Timeline delays without notification
- Limited post-coordination support
- Poor API implementation
- Inadequate training resources

Case Example: Vendor Reality Check

Community Health Network's EHR vendor assured "full ICD-11 readiness." Three months before go-live, testing revealed:

- Post-coordination non-functional
- API connections failed security requirements
- Mapping tables contained significant errors
- Training materials were generic WHO content

Recovery required:

- Third-party coding tool integration

- Custom interface development
- Internal mapping validation
- Development of proprietary training

Vendor Management Strategy:

- Require detailed functionality demonstrations
- Conduct proof-of-concept testing
- Include penalties in contracts
- Maintain vendor-agnostic contingencies
- Collaborate with other clients

8. Inadequate Testing Strategies

Superficial testing misses critical failures that emerge under real-world conditions.

Testing Shortcuts That Backfire:

- Using simple cases only
- Testing individual systems in isolation
- Ignoring performance under load
- Missing edge cases
- Skipping end-to-end scenarios

Comprehensive Testing Protocol:

Unit Testing: Each component individually

- Code search functionality
- Post-coordination validation
- Mapping accuracy
- Display formatting

Integration Testing: Connected systems

- EHR to billing system
- Quality reporting pulls

- Research database feeds
- External interfaces

User Acceptance Testing: Real-world scenarios

- Common diagnoses by department
- Complex multi-condition patients
- Workflow completion times
- Error handling

Performance Testing: System limits

- Concurrent user loads
- Response times
- Database query efficiency
- Network bandwidth usage

9. Neglecting Change Management

Technical implementation without cultural change guarantees suboptimal outcomes.

Change Management Gaps:

- No burning platform for change
- Lack of visible leadership support
- Inadequate recognition programs
- Missing feedback mechanisms
- Absent course correction processes

Successful Change Management Elements:

Creating Urgency:

- Share competitive disadvantages
- Highlight quality improvement opportunities
- Demonstrate efficiency gains
- Show patient care benefits

Building Coalition:

- Physician champions in each specialty
- Nursing leadership engagement
- Administrative support visible
- Patient story integration

Sustaining Change:

- Regular optimization cycles
- Celebrating successes publicly
- Addressing concerns transparently
- Continuous improvement mindset

10. Insufficient Quality Monitoring

Organizations often discover problems through denied claims or failed audits rather than proactive monitoring.

Quality Blind Spots:

- Assuming accuracy from day one
- Monitoring only coding productivity
- Ignoring clinical documentation quality
- Missing relationship coding patterns
- Delayed financial impact assessment

Proactive Quality Framework:

Daily Monitoring:

- Coding accuracy spot checks
- System performance metrics
- User support ticket patterns
- Error message frequencies

Weekly Analysis:

- Departmental coding patterns
- Documentation completeness
- Claim acceptance rates
- Query patterns

Monthly Review:

- Comprehensive quality audits
- Financial impact assessment
- Stakeholder satisfaction surveys
- Optimization opportunity identification

Technical Integration Challenges

Beyond organizational pitfalls, technical challenges require specific attention.

API Implementation Complexities

Modern architectures assume API connectivity, but implementation proves challenging:

- Firewall configurations blocking connections
- Authentication protocol mismatches
- Version control synchronization
- Performance degradation under load

Solution Approach:

- Early security team engagement
- Comprehensive API testing
- Redundancy planning
- Performance benchmarking

Data Migration Pitfalls

Historical data mapping seems straightforward until attempted at scale:

- One-to-many mappings requiring clinical decisions
- Temporal data losing granularity
- Relationship information absent in legacy systems
- Volume overwhelming manual review

Migration Strategy:

- Automated mapping for high-confidence codes
- Clinical review for complex cases
- Acceptance of some historical data limitations
- Clear documentation of mapping decisions

Quality Assurance Strategies

Preventing pitfalls requires robust quality assurance throughout implementation and beyond.

Pre-Implementation Quality:

- Validated mapping tables
- Tested workflows
- Competency-based training completion
- System performance benchmarks

Implementation Quality:

- Real-time monitoring dashboards
- Rapid issue identification
- Clear escalation pathways
- Daily optimization cycles

Post-Implementation Quality:

- Continuous improvement processes
- Regular audit programs
- Stakeholder feedback integration
- Emerging issue identification

These pitfalls represent real pain experienced by organizations worldwide. Each mistake costs time, money, and organizational trust. By understanding these patterns, you can build implementation plans that navigate around known obstacles.

Success requires humility—acknowledging the complexity and learning from others' experiences. No implementation proceeds perfectly, but avoiding major pitfalls enables manageable challenges rather than crises.

Key Takeaways

- Treat ICD-11 as organizational transformation, not technical upgrade
- Realistic timelines span 18-48 months for most organizations
- Stakeholder engagement must include all affected groups, not just obvious users
- Training investment requires conceptual understanding beyond technical skills
- Workflows must evolve to leverage ICD-11 capabilities
- Communication strategies need audience-specific messaging through multiple channels
- Vendor readiness requires verification, not trust
- Testing must include real-world complexity and edge cases
- Change management ensures cultural adoption beyond technical implementation
- Quality monitoring must be proactive and comprehensive

Chapter 11: Technology and Software Considerations

The digital architecture of ICD-11 demands thoughtful technology decisions. Unlike previous classification systems that functioned adequately with basic database storage, ICD-11 requires modern infrastructure, sophisticated integration, and careful vendor selection. Technology choices made today determine implementation success and long-term sustainability.

System Requirements and Infrastructure

Understanding ICD-11's technical demands prevents infrastructure surprises during implementation. Requirements extend beyond simple storage and retrieval to support real-time validation, complex queries, and seamless integration.

Core Infrastructure Requirements

Modern systems need capabilities unimaginable when ICD-10 launched:

Processing Power: Post-coordination validation happens in milliseconds during clinical documentation. A single code entry might trigger:

- Syntax validation
- Semantic relationship checking
- Clinical rule evaluation
- Duplicate detection
- Suggestion generation

Dr. Kevin Liu, Chief Technology Officer at a large health system, explains the impact: "ICD-10 queries were simple lookups. ICD-11 requires graph traversal through semantic relationships. Our database servers needed 3x processing power to maintain response times" (41).

Memory Requirements: The semantic foundation loads into memory for performance:

- Foundation component: 2-4 GB minimum
- Linearization tables: 500 MB per language
- Relationship graphs: 1-2 GB
- Validation rules: 500 MB
- Cache requirements: Variable by usage

Storage Considerations: Beyond raw data, consider:

- Multiple language versions
- Historical mappings for migration
- Audit trails for all changes
- Version control for updates
- Backup redundancy

Network Architecture: API-based architecture demands robust networking:

- Bandwidth for real-time API calls
- Low latency for clinical responsiveness
- Redundant paths for reliability
- Security protocols for external connections
- Quality of Service prioritization

Case Example 1: Infrastructure Scaling

Regional Medical Center initially underestimated requirements. Their go-live experienced:

- 30-second delays for complex post-coordination
- System timeouts during peak documentation
- Report generation failures
- User frustration and resistance

Emergency infrastructure upgrades included:

- Database server CPU cores doubled
- Memory increased from 32GB to 128GB
- Solid-state storage for frequent queries
- Network optimization for API traffic
- Caching layer implementation

Post-upgrade performance exceeded ICD-10 baselines, validating adequate infrastructure investment.

Security Architecture

ICD-11's external connectivity introduces security considerations:

API Security:

- OAuth 2.0 authentication minimum
- Token management systems
- API gateway implementation
- Rate limiting to prevent abuse
- Encryption for all transmissions

Access Control:

- Role-based permissions
- Audit trails for all access
- Multi-factor authentication
- Session management
- Principle of least privilege

Data Protection:

- Encryption at rest and in transit
- Backup security
- Disaster recovery planning
- Business continuity provisions
- Compliance with regulations (HIPAA, GDPR)

EHR/EMR Integration Guidelines

Electronic health record integration determines clinical usability. Poor integration frustrates users and limits ICD-11's benefits.

Integration Architecture Patterns

Native Integration: EHR vendor provides built-in ICD-11 support

- Advantages: Seamless workflow, vendor support, unified interface
- Disadvantages: Dependent on vendor timeline, limited customization
- Best for: Organizations with modern EHRs committed to ICD-11

Bolt-On Integration: Third-party coding tool interfaces with EHR

- Advantages: Best-of-breed selection, faster availability, specialized features
- Disadvantages: Interface complexity, potential workflow disruption
- Best for: Organizations needing advanced features or early adoption

Hybrid Approach: Native basic functions with specialized add-ons

- Advantages: Balanced functionality and integration
- Disadvantages: Multiple vendor management
- Best for: Large organizations with diverse needs

Case Example 2: Integration Evolution

Academic Medical Center's integration journey:

Phase 1: Waited for EHR vendor's native ICD-11

- 18-month delay from promised date
- Basic functionality when delivered
- No post-coordination support

Phase 2: Added third-party coding assistant

- Immediate advanced capabilities
- Required separate login initially
- Workflow disruption complaints

Phase 3: Achieved single sign-on integration

- Seamless user experience
- Combined best features
- High user satisfaction

Technical Integration Requirements

Data Exchange Standards:

- FHIR R4 for terminology services
- HL7 for legacy system compatibility
- JSON for modern API communication
- XML for structured documents

Real-Time Requirements:

- Sub-second response for code lookup
- Immediate validation feedback
- Background processing for complex operations
- Asynchronous options for batch operations

Workflow Integration Points:

Problem List Management:

- Automatic ICD-11 coding from clinical terms
- Post-coordination suggestions based on context
- Historical code migration
- Relationship preservation

Clinical Documentation:

- Natural language processing assists
- Real-time coding suggestions
- Validation during entry
- Query prevention through prompts

Order Entry Integration:

- Diagnosis-driven order sets
- Insurance pre-authorization codes
- Clinical decision support triggers
- Quality measure identification

Integration Testing Essentials

Comprehensive testing prevents go-live failures:

Functional Testing:

- Code search accuracy
- Post-coordination creation
- Validation rule enforcement
- Error message clarity

Performance Testing:

- Response times under load
- Concurrent user limits
- Database query optimization
- Network bandwidth utilization

Workflow Testing:

- End-to-end scenario completion
- User acceptance validation
- Time-motion studies
- Error recovery procedures

API Implementation

APIs transform ICD-11 from static classification to dynamic service. Proper implementation enables real-time updates, seamless integration, and future flexibility.

API Architecture Decisions

Direct WHO API Connection:

- Advantages: Always current, authoritative source
- Disadvantages: Internet dependency, potential latency
- Governance needs: Change management for updates
- Best for: Organizations prioritizing currency

Local API Cache/Proxy:

- Advantages: Performance control, offline capability
- Disadvantages: Synchronization complexity, storage needs
- Governance needs: Update procedures, version control
- Best for: High-volume or reliability-critical environments

Case Example 3: API Strategy Success

Multi-hospital system implemented tiered API architecture:

Tier 1: Local cache for common codes (90% of queries)

- Sub-100ms response times
- Updated nightly
- 50,000 most-used codes

Tier 2: Regional cache for extended content

- 200ms average response
- Updated weekly
- Complete national linearization

Tier 3: Direct WHO API for rare queries

- 500ms-2s response
- Real-time updates
- Rarely accessed codes

Results:

- 95% queries served from Tier 1
- System resilience during internet outages
- Optimal performance/currency balance

API Implementation Best Practices

Authentication and Security:

```
- Implement OAuth 2.0 flows
- Rotate tokens regularly
- Monitor for unusual patterns
- Log all API access
- Encrypt sensitive parameters
```

Error Handling:

```
- Graceful degradation for API failures
- User-friendly error messages
- Automatic retry logic
- Circuit breaker patterns
- Fallback to cached data
```

Performance Optimization:

```
- Connection pooling
- Request batching where possible
- Compression for large responses
- Pagination for result sets
- Caching strategies by data type
```

Monitoring and Analytics:

```
- Response time tracking
- Error rate monitoring
- Usage pattern analysis
```

```
- Capacity planning metrics
- Cost optimization (if metered)
```

Software Selection Guide

Choosing the right software stack determines implementation success. The market offers diverse options, each with strengths and limitations.

Evaluation Framework

Core Functionality Assessment:

Essential Features:

- Complete ICD-11 linearization support
- Post-coordination creation and validation
- Multi-language capability
- API integration options
- Mapping tools for migration

Advanced Features:

- Natural language processing
- Machine learning suggestions
- Clinical decision support integration
- Analytics and reporting
- Workflow automation

Vendor Evaluation Criteria

Technical Capabilities:

- Architecture modernity
- Scalability provisions
- Integration flexibility
- Performance benchmarks
- Security certifications

Organizational Factors:

- Company stability
- Implementation experience
- Support quality
- Training resources
- User community size

Financial Considerations:

- Licensing models
- Implementation costs
- Ongoing maintenance
- Scaling pricing
- Hidden expenses

Case Example 4: Selection Process Excellence

Large health system's systematic selection:

Phase 1: Requirements Definition

- 127 functional requirements identified
- Weighted by criticality
- Validated with stakeholders

Phase 2: Market Survey

- 12 vendors identified
- 7 met minimum requirements
- 4 invited for detailed evaluation

Phase 3: Proof of Concept

- Real scenarios tested
- Performance benchmarked
- Integration complexity assessed
- User feedback gathered

Phase 4: Selection Decision

- Scoring matrix applied
- Reference checks conducted
- Negotiation completed
- Implementation planned

Common Software Categories

Enterprise EHR Modules:

- Examples: Epic, Cerner, Athenahealth modules
- Pros: Integrated workflow, vendor support
- Cons: Limited flexibility, release dependent
- Best fit: Committed EHR users wanting simplicity

Specialized Coding Platforms:

- Examples: 3M CodeFinder, Optum CAC
- Pros: Advanced features, coding expertise
- Cons: Integration requirements, additional cost
- Best fit: High-volume coding operations

Terminology Servers:

- Examples: Open-source HAPI FHIR, commercial solutions
- Pros: Standards-based, flexible deployment
- Cons: Technical expertise required
- Best fit: Organizations with strong IT

Analytics Platforms:

- Examples: Business intelligence tools with ICD-11
- Pros: Powerful analysis capabilities
- Cons: May require data transformation
- Best fit: Research-focused organizations

Technology decisions shape every aspect of ICD-11 implementation. Infrastructure must support not just current needs but future growth. Integration architecture determines clinical usability. API implementation enables the dynamic capabilities that differentiate ICD-11. Software selection commits organizations to specific capabilities and constraints.

Approach these decisions methodically. Engage technical and clinical stakeholders throughout. Plan for growth and change. Most importantly, remember that technology serves clinical care—every decision should improve patient outcomes.

Key Takeaways

- Infrastructure requirements include 3x processing power and sophisticated networking versus ICD-10
- Security architecture must address API connectivity and external integration risks
- EHR integration patterns include native, bolt-on, and hybrid approaches
- API implementation benefits from tiered architecture balancing performance and currency
- Comprehensive testing must include functional, performance, and workflow validation
- Software selection requires systematic evaluation of functionality, vendor, and financial factors
- Local caching strategies optimize performance while maintaining accuracy
- Integration points span problem lists, documentation, orders, and analytics
- Future flexibility should guide current technology decisions

Chapter 12: Training Resources and Certification Guidance

Professional excellence in ICD-11 requires more than memorizing codes—it demands understanding classification philosophy, mastering technical skills, and maintaining ongoing competency. The global transition creates unprecedented demand for qualified professionals, making systematic training and certification essential for career advancement and organizational success.

WHO Official Training Programs

The World Health Organization provides foundational training resources, setting global standards for ICD-11 education. Understanding these offerings helps organizations build comprehensive training strategies.

WHO Academy Core Curriculum

The WHO Academy offers structured learning paths:

Foundation Course (10 hours):

- ICD-11 philosophy and structure
- Navigation fundamentals
- Basic coding principles
- Simple post-coordination
- Assessment with certificate

Dr. Linda Peterson, who coordinates training for a health system, observes: "The WHO Foundation Course provides excellent conceptual grounding. Every trainee completes it before our hands-on sessions. It establishes common vocabulary and understanding" (42).

Specialized Modules:

- Mortality coding (8 hours)
- Morbidity applications (12 hours)
- Primary care subset (6 hours)
- Traditional medicine (4 hours)
- Implementation planning (8 hours)

Language Availability: Currently offered in 14 languages with automatic certificate generation in learner's chosen language. New translations release quarterly.

Case Example 1: National Training Strategy

Thailand's Ministry of Health built upon WHO resources:

Phase 1: All trainers completed WHO Academy courses

- 200 master trainers certified
- Localization rights obtained
- Thai examples added

Phase 2: Cascaded training nationwide

- WHO content as foundation
- Local clinical scenarios
- Mobile-friendly delivery
- Social media support groups

Phase 3: Ongoing competency maintenance

- Monthly webinars
- Peer learning networks
- Annual assessments
- WHO updates integrated

Results: 15,000 healthcare workers trained in 18 months with 88% passing competency assessments.

WHO Reference Materials

Beyond courses, WHO provides:

Reference Guide: Comprehensive 400+ page manual

- Detailed classification rules
- Post-coordination guidance
- Specialty-specific sections
- Example libraries

Browser Training Mode: Practice environment

- Safe exploration space
- Tutorial overlays
- Guided exercises
- No production impact

Coding Tools: Interactive utilities

- Post-coordination builder
- Validation checker
- Mapping assistant
- Code search optimization

Implementation Toolkit: Project resources

- Planning templates
- Communication materials
- Assessment tools
- Quality frameworks

Professional Certification Pathways

Certification validates competency and advances careers. Multiple organizations offer credentials, each with distinct focus and recognition.

American Health Information Management Association (AHIMA)

AHIMA leads US certification efforts:

Certified ICD-11 Coder (CIC-11):

- Prerequisites: Active CCS or CCS-P
- Exam: 4 hours, 150 questions
- Domains: Classification principles, post-coordination, implementation
- Passing score: 70%
- Renewal: 20 CEUs biennially

Specialty Certifications:

- ICD-11 Trainer (CIC-11-T)
- Implementation Specialist (CIC-11-I)
- Auditor Certification (CIC-11-A)

Case Example 2: Career Advancement Through Certification

Jennifer Martinez, veteran coder, pursued systematic certification:

Starting point: 15 years ICD-10 experience, CCS credential

Year 1:

- Completed WHO Academy courses
- Attended AHIMA boot camp
- Passed CIC-11 exam
- Promoted to lead coder

Year 2:

- Earned trainer certification
- Led organizational training
- Consulted for other facilities
- 40% salary increase

Year 3:

- Implementation specialist credential
- Director of coding position
- Regional expert recognition
- Speaking at conferences

"Certification opened doors I didn't know existed. Organizations desperately need ICD-11 expertise" (43).

International Credentials

WHO-FIC Certification:

- Global recognition
- Competency-based assessment
- Multiple specialty tracks
- Annual maintenance requirements

Country-Specific Programs:

- Australia: HIMAA certification
- Canada: CHIMA credentials
- UK: NHS Digital programs
- Germany: DIMDI certification

Each reflects national implementation approaches while maintaining international standards.

Specialty Society Certifications

Medical specialties develop focused credentials:

Oncology: Cancer registry ICD-11 certification **Cardiology**: Cardiovascular coding specialization **Mental Health**: Behavioral health ICD-11 credential **Surgery**: Procedural coding with ICD-11 diagnoses

These complement general certifications with specialty depth.

Creating Your Training Plan

Effective training plans balance individual needs, organizational requirements, and resource constraints. Systematic approaches yield better outcomes than ad hoc training.

Organizational Needs Assessment

Begin with comprehensive analysis:

Role Mapping:

- Who uses ICD codes?
- What level of detail needed?
- When in workflows?
- Why (purpose of usage)?

Skill Gap Analysis:

- Current ICD-10 proficiency
- Technology comfort levels
- Change readiness
- Learning preferences

Volume Calculations:

- Number requiring training
- Geographic distribution
- Schedule constraints
- Language requirements

Training Modality Selection

Match delivery methods to needs:

Instructor-Led Training:

- Best for: Complex concepts, hands-on practice

- Advantages: Immediate feedback, peer learning
- Disadvantages: Scheduling challenges, cost
- Optimal class size: 12-20 participants

E-Learning Modules:

- Best for: Foundation concepts, self-paced learning
- Advantages: Scalability, consistency, flexibility
- Disadvantages: Limited interaction, requires discipline
- Completion rates: 60-70% without accountability

Blended Learning:

- Best for: Comprehensive programs
- Combines: Online foundations with classroom application
- Advantages: Flexibility with interaction
- Typical structure: 60% online, 40% classroom

Case Example 3: Multi-Modal Success

Regional health network's training approach:

Foundation Phase (Months 1-2):

- WHO Academy courses (online, self-paced)
- Weekly virtual office hours
- Discussion forums
- Pre-assessment testing

Application Phase (Months 3-4):

- Role-specific classroom sessions
- Department scenarios
- Hands-on system practice
- Peer mentoring

Reinforcement Phase (Months 5-6):

- Monthly refresher webinars
- Quick reference tools deployment
- Super-user networks
- Real-case reviews

Sustainment Phase (Ongoing):

- Quarterly updates training
- Annual competency validation
- Advanced feature workshops
- Optimization sessions

Results: 95% competency achievement, 89% user satisfaction, 12% productivity improvement post-training.

Curriculum Development Best Practices

Adult Learning Principles:

- Problem-centered, not subject-centered
- Experience-based examples
- Immediate applicability
- Self-directed options
- Clear relevance to roles

Content Sequencing:

1. Why ICD-11 matters (motivation)
2. Core concepts (foundation)
3. Basic skills (competence)
4. Advanced features (mastery)
5. Optimization techniques (excellence)

Practice Integration:

- 20% lecture/demonstration
- 40% guided practice
- 30% independent exercises

- 10% assessment/feedback

Dr. Robert Chen, medical educator, emphasizes: "Adults learn by doing. Every concept needs immediate practice with relevant examples. Generic exercises frustrate experienced professionals—use their actual work scenarios" (44).

Training Material Development

Quick Reference Guides:

- Role-specific cards
- Common code listings
- Post-coordination examples
- Workflow diagrams
- Error resolution steps

Practice Datasets:

- Department-specific cases
- Progressive complexity
- Answer keys with rationale
- Common error examples
- Edge case scenarios

Assessment Tools:

- Pre-training knowledge checks
- Module quizzes
- Practical exercises
- Competency demonstrations
- Certification preparation

Competency Assessment Methods

Measuring training effectiveness ensures capability, not just completion. Multi-dimensional assessment provides comprehensive validation.

Knowledge Assessment

Traditional testing measures understanding:

Multiple Choice Questions:

- Tests conceptual knowledge
- Efficient for large groups
- Objective scoring
- Limited practical validation

Example: "Post-coordination in ICD-11 uses which symbol to link stem codes with extension codes? a) Forward slash (/) b) Ampersand (&) c) Plus sign (+) d) Hyphen (-)"

Case-Based Questions:

- Applies knowledge to scenarios
- Tests judgment and reasoning
- More realistic validation
- Time-intensive development

Practical Skills Assessment

Real-world capability matters most:

Coding Exercises:

- Actual chart documentation
- Time-limited scenarios
- Accuracy measurement
- Efficiency tracking

Workflow Simulations:

- End-to-end processes
- System navigation
- Error recognition

- Problem resolution

Quality Audits:

- Production work review
- Accuracy rates
- Completeness assessment
- Pattern analysis

Case Example 4: Comprehensive Competency Program

Academic medical center's assessment approach:

Initial Certification:

- Knowledge exam (100 questions, 80% passing)
- Practical coding (20 cases, 90% accuracy)
- Workflow completion (5 scenarios)
- Peer review validation

Ongoing Validation:

- Monthly accuracy audits
- Quarterly knowledge updates
- Annual recertification
- Continuous education credits

Advanced Competencies:

- Post-coordination mastery
- Complex case management
- Teaching qualification
- Quality auditor certification

Results: Clear competency standards improved coding accuracy 25% and reduced training time for new hires 30%.

Behavioral Competencies

Beyond technical skills:

Change Adaptability:

- Openness to new methods
- Continuous learning mindset
- Flexibility with ambiguity
- Innovation adoption

Collaboration Skills:

- Clinical communication
- Query effectiveness
- Team problem-solving
- Knowledge sharing

Critical Thinking:

- Pattern recognition
- Problem analysis
- Decision-making
- Quality focus

Performance Metrics Framework

Comprehensive measurement includes:

Individual Metrics:

- Coding accuracy rates
- Productivity levels
- Error patterns
- Learning progression

Team Metrics:

- Department accuracy
- Peer support levels

- Knowledge sharing
- Collective improvement

Organizational Metrics:

- Overall accuracy trends
- Financial impact
- Quality outcomes
- Implementation progress

Continuous Improvement Integration

Assessment drives improvement:

Individual Development Plans:

- Strength recognition
- Gap identification
- Learning paths
- Progress tracking

Targeted Interventions:

- Remedial training
- Mentoring programs
- Practice opportunities
- Resource provision

System Optimization:

- Common error analysis
- Workflow refinement
- Tool enhancement
- Process improvement

ICD-11 success depends on well-trained, certified professionals. Organizations investing in comprehensive training and certification reap benefits through improved accuracy, efficiency, and staff retention. Individual professionals advancing their skills find expanding career opportunities in a rapidly evolving field.

The journey from novice to expert requires commitment, but established pathways ease the transition. Whether pursuing WHO foundations, professional certifications, or organizational excellence, systematic approaches yield superior outcomes.

Remember: ICD-11 represents not just new codes but new thinking about health information. Training and certification that embrace this philosophy prepare professionals for current needs and future innovations.

Key Takeaways

- WHO Academy provides free, standardized foundation training in 14 languages
- Professional certifications from AHIMA and international bodies validate competency
- Specialty certifications add depth for focused practice areas
- Effective training plans use blended learning matching adult learning principles
- Role-specific curricula improve relevance and engagement
- Competency assessment must include knowledge, practical skills, and behaviors
- Multi-dimensional metrics track individual, team, and organizational performance
- Continuous improvement cycles optimize training effectiveness
- Investment in training and certification yields career advancement and organizational success
- Success requires embracing ICD-11 philosophy, not just memorizing codes

Chapter 13: Quick Reference Guides and Cheat Sheets

The moment arrives when you need that critical code during a patient encounter. Your mind goes blank. The computer freezes. Time ticks away. This chapter provides the essential references you'll reach for repeatedly—dog-eared pages, bookmarked screens, laminated cards tucked in your pocket. These aren't just lists; they're lifelines crafted from real-world experience.

ICD-11 Code Structure Quick Reference

Understanding ICD-11's architecture starts with recognizing its elegant simplicity. Unlike ICD-10's variable-length codes that seemed designed by committee, ICD-11 follows a consistent pattern that makes sense once you grasp the logic.

Every ICD-11 code begins with a number or letter indicating the chapter. The second character is always a letter—but never I or O, preventing confusion with numbers 1 and 0. The third and fourth positions are always numbers. This creates the stem code foundation: one character for chapter, followed by letter-number-number.

For example, **5A11** represents Type 2 diabetes mellitus. The 5 indicates Chapter 5 (Endocrine, nutritional or metabolic diseases), A is the block identifier, and 11 specifies the particular condition. Simple. Predictable. Logical.

Extensions follow a decimal point, adding specificity without changing the core meaning. Think of extensions as adjectives modifying a noun—they describe but don't transform. The code 5A11.0 still means Type 2 diabetes, just with the added detail of poor blood glucose control.

Case Example 1: Emergency Department Coding

Dr. Sarah Mitchell works the night shift when a patient arrives with acute chest pain. Under pressure, she needs to code quickly and accurately. Her reference card shows:

- **MG30** represents the stem for chest pain
- Add **.0Y** for chest pain, unspecified
- Post-coordinate with temporal markers if needed

She codes MG30.0Y for the initial presentation. Later, when tests reveal gastroesophageal reflux as the cause, she can update to show the relationship between the symptom and final diagnosis using post-coordination. The structure's consistency means she doesn't waste time hunting through multiple code formats.

The beauty lies in scalability. Need more detail? Add extension codes. Need to show relationships? Use post-coordination. But the basic four-character stem remains your anchor—always there, always structured the same way.

Post-coordination Rules

Post-coordination transforms ICD-11 from a list of codes into a language for describing clinical reality. Two symbols create all the magic: the forward slash (/) and the ampersand (&). Master these, and you master ICD-11.

The forward slash (/) links clinical concepts with relationships. It says "this connects to that"—not just listing two things, but showing how they relate. Use it for:

- Manifestations of underlying diseases
- Causes and their effects
- Primary conditions and their complications

The ampersand (&) adds descriptive details to a stem code. It's like adding multiple adjectives to describe one noun. Use it for:

132

- Anatomical locations
- Severity indicators
- Temporal qualifiers
- Any extension that modifies the stem

Case Example 2: Complex Chronic Disease

Maria Rodriguez, a family physician, treats John Patterson, age 67, with multiple chronic conditions. His diabetes has led to kidney disease and retinopathy. Using her post-coordination reference:

Primary condition: **5A11** (Type 2 diabetes mellitus)

- Forward slash to show consequences: **/5A25.1** (Diabetic kidney disease, stage 3)
- Another forward slash for additional complication: **/9B71.02** (Diabetic retinopathy, proliferative)
- Ampersand for control status: **&XN3H5** (HbA1c 8.5%)

The complete code cluster: **5A11/5A25.1/9B71.02&XN3H5**

This single expression captures what previously required four separate codes with unclear relationships. The forward slashes show diabetes caused both complications. The ampersand adds the crucial control metric. One cluster tells the complete clinical story.

Critical rules to follow:

1. **Stem codes come first**—always start with the primary condition
2. **Order matters**—list complications by clinical significance
3. **Validate combinations**—not every code can link with every other code
4. **Use appropriate symbols**—forward slash for relationships, ampersand for modifications
5. **Keep it clinically logical**—if it doesn't make medical sense, it's probably wrong

Common Code Conversions

The transition from ICD-10 to ICD-11 requires a reliable conversion reference. While automated mapping tools exist, understanding common conversions helps you code confidently and catch mapping errors.

High-frequency conversions every coder needs:

Diabetes codes undergo significant reorganization. ICD-10's E10-E14 series, with their complex fourth and fifth characters for complications, simplify in ICD-11:

- E10.9 (Type 1 diabetes without complications) becomes **5A10.Z**
- E11.65 (Type 2 diabetes with hyperglycemia) becomes **5A11.0**
- E11.21 (Type 2 with diabetic nephropathy) becomes **5A11/5A25** with appropriate staging

Hypertension codes reflect modern understanding of the condition:

- I10 (Essential hypertension) maps to **BA00.Z**
- I11.9 (Hypertensive heart disease) becomes **BA01** with specific cardiac manifestations post-coordinated
- I12.9 (Hypertensive chronic kidney disease) converts to **BA02** with kidney disease stage specified

Mental health codes show the most dramatic changes:

- F32.0 (Mild depressive episode) becomes **6A70.0** (Single episode depressive disorder, mild)
- F84.0 (Childhood autism) transforms to **6A02** (Autism spectrum disorder) with specifiers
- F41.1 (Generalized anxiety disorder) maps to **6B73** with new severity scales

Case Example 3: Outpatient Clinic Conversion

Dr. James Chen reviews his clinic's most common diagnoses before ICD-11 go-live. His nurse prepared a conversion sheet:

Mrs. Thompson's chronic conditions:

- **Old**: E11.9, I10, F32.1, M79.3
- **New**: 5A11.Z (diabetes, control unspecified—prompting better documentation)
- BA00.Z (hypertension—but now he can add blood pressure readings)
- 6A70.1 (moderate depression—with new function impact scales)
- MG70.3 (fibromyalgia—finally properly classified, not "miscellaneous")

The conversion reveals documentation improvement opportunities. ICD-11 doesn't just translate old codes—it pushes toward better clinical specificity.

Infection codes streamline while adding precision:

- J18.9 (Pneumonia, unspecified) becomes **CA40.Z** with required organism specification when known
- N39.0 (Urinary tract infection) transforms to **GC08** with anatomical location options
- A41.9 (Sepsis, unspecified) converts to **1G43** with source identification expected

Decision Trees and Flowcharts

Visual guides speed decision-making when choosing between similar codes or determining post-coordination needs. These flowcharts distill complex logic into simple pathways.

Primary Diagnosis Selection Flow:

Start with the presenting problem. Ask: Is this a symptom or confirmed diagnosis?

- If symptom only → Use symptom code, prepare to update later
- If confirmed diagnosis → Proceed to condition coding

Next, determine: Is this condition primary or secondary?

- If primary → Code as stem
- If secondary to another condition → Consider post-coordination with cause

Then assess: Are there active complications?

- If yes → Add via forward slash
- If no → Consider severity or control status extensions

Post-coordination Decision Tree:

Begin with your stem code selected. Ask: Does this condition have a known cause?

- If yes → Add causing condition with forward slash
- If no → Proceed to manifestations

Check: Are there current manifestations or complications?

- If yes → Add each with forward slash in order of clinical importance
- If no → Consider descriptive extensions

Finally: What details improve clinical understanding?

- Anatomical location → Add with ampersand
- Severity indicators → Add with ampersand
- Temporal markers → Add with ampersand

Emergency Severity Coding Flow:

Patient presents to emergency department. First decision: Life-threatening?

- If yes → Code primary threat first, add contributing factors
- If no → Code by chief complaint, then findings

Assessment complete. Next question: Single or multiple issues?

- If single → Standard coding with appropriate extensions
- If multiple → Determine relationships for clustering

Disposition decided. Final consideration: Admission or discharge?

- If admission → Include severity markers supporting decision
- If discharge → Document resolution or stability codes

These visual tools transform complex decisions into manageable steps. Post them where you code. Share them with new staff. Update them as you discover patterns in your practice.

Quick references work only if they're quick to access. Print these guides. Laminate them. Pin them up. Save them on your phone. Make them part of your coding environment. The investment in creating personal quick references pays dividends every day—in time saved, errors prevented, and confidence maintained.

Key Takeaways

- ICD-11's four-character stem structure (Chapter-Letter-Number-Number) provides consistent code foundation
- Post-coordination uses forward slash (/) for relationships and ampersand (&) for descriptive extensions
- Common conversions from ICD-10 to ICD-11 often reveal documentation improvement opportunities
- High-frequency codes in diabetes, hypertension, and mental health show significant mapping changes
- Decision trees simplify complex coding choices into step-by-step logic
- Visual flowcharts for diagnosis selection, post-coordination, and emergency coding speed decision-making

- Personal quick references should be accessible in multiple formats for different coding environments
- Regular use of reference tools builds pattern recognition and coding confidence

Chapter 14: Glossary of Terms

Medical classification speaks its own language. New terms emerge with ICD-11 that never existed before. Old terms take on new meanings. This glossary serves as your translator, turning technical terminology into practical understanding. Each definition includes not just what terms mean, but why they matter to your daily practice.

Core ICD-11 Terminology

Foundation Component The complete knowledge base of ICD-11 containing over 85,000 unique entities with all their relationships, definitions, and properties. Think of it as the master library holding everything known about diseases and health conditions. Unlike the linear lists you use for coding, the Foundation allows multiple parents for each entity—mirroring how diseases actually work. A condition like diabetic retinopathy lives under both diabetes and eye diseases in the Foundation, though you'll code it in just one place.

Linearization A practical subset of the Foundation organized for specific use. If the Foundation is the complete library, linearizations are the reference books you actually pull off the shelf. The Mortality and Morbidity Statistics (MMS) linearization contains about 35,000 codes organized hierarchically for statistical reporting. Other linearizations exist for primary care (simplified) or research (detailed). Each linearization maintains links back to the Foundation, ensuring consistency while meeting different needs.

Stem Code The four-character alphanumeric code forming the basis of every ICD-11 diagnosis. Format follows Chapter-Letter-Number-Number pattern, like 5A11 for Type 2 diabetes. Stem codes stand alone as complete diagnoses but serve as launching points for adding detail through post-coordination. Every valid code starts with a stem—no exceptions.

Extension Code Additional codes that cannot stand alone but add detail to stem codes. These X-codes specify anatomy, severity, temporality, and other clinical details. For instance, XK9J indicates "right side" and combines with fracture codes to specify laterality. Extensions multiply the expressiveness of stem codes without creating endless pre-combined options.

Case Example 1: Foundation to Practice

Dr. Kim needs to code a patient with lung cancer metastatic to bone. In the Foundation Component, this concept has multiple parents—it's both a lung condition and a bone condition. But in the MMS linearization she uses for coding, she finds:

- Stem code: **2C25.1** (Adenocarcinoma of bronchus or lung)
- Post-coordinated with: **/2D12.3** (Secondary malignant neoplasm of bone)

The Foundation's rich relationships inform the linearization's practical structure. Understanding this connection helps her appreciate why post-coordination works as it does.

Post-coordination The ability to combine codes to create detailed clinical descriptions. Using forward slashes (/) and ampersands (&), post-coordination builds precise diagnostic statements from simpler elements. This replaces ICD-10's attempt to pre-combine every possible variation. Post-coordination turns coding from selecting preset options to constructing accurate clinical descriptions.

Cluster Coding Grouping related codes together to describe complex clinical scenarios. A cluster maintains relationships between conditions—showing what caused what, what manifests with what, what complicates what. Proper clustering tells clinical stories rather than listing disconnected diagnoses.

URI (Uniform Resource Identifier) A permanent, unique identifier for each ICD-11 entity. Like a social security number for diseases, the URI never changes even as classifications update. This enables computer systems to track concepts across versions and languages. For

example, diabetes mellitus has URI
http://id.who.int/icd/entity/1271944134 regardless of code changes.

Content Model The structured template defining what information belongs with each ICD-11 entity. It specifies what details can be recorded—definitions, synonyms, diagnostic criteria, exclusions. The Content Model ensures consistency across the entire classification. Understanding it helps you predict what information you'll find for any code.

Technical Architecture Terms

API (Application Programming Interface) The communication protocol allowing computer systems to interact with ICD-11. Instead of downloading static files, systems query the API for current codes, validate combinations, and retrieve updates. APIs make ICD-11 a living service rather than a fixed publication.

FHIR (Fast Healthcare Interoperability Resources) The modern standard for healthcare data exchange. ICD-11's FHIR compatibility means diagnoses flow seamlessly between systems without translation. As healthcare adopts FHIR broadly, ICD-11 codes travel unchanged across platforms.

Case Example 2: API in Action

Community Hospital's EHR connects to ICD-11's API. When Dr. Patel types "chest pain" into the diagnosis field:

1. The EHR sends this text to the API
2. API returns relevant codes with descriptions
3. Dr. Patel selects **MG30.0Y** (Chest pain, unspecified)
4. API validates this is a proper stem code
5. API suggests relevant extensions for location, severity
6. Final selection returns to EHR with full code details

This happens in milliseconds, feeling like the codes live in the EHR itself. The API makes ICD-11 dynamic and always current.

Semantic Web The network of meaning connecting ICD-11 entities. Beyond simple hierarchies, the semantic web captures rich relationships—causal, temporal, anatomical. This enables intelligent queries like "find all conditions that can cause kidney failure" returning results across multiple chapters.

JSON (JavaScript Object Notation) A data format for exchanging ICD-11 information between systems. Lightweight and readable by both humans and machines, JSON carries code details, relationships, and metadata efficiently. Most modern systems speak JSON natively.

Postcoordination Validation Automated checking ensuring code combinations make clinical sense. The system prevents impossible combinations like pregnancy in males or fractures of organs without bones. Validation rules come from medical knowledge encoded in the Foundation.

Implementation Terms

Dual Coding Period The transition phase where organizations code in both ICD-10 and ICD-11 simultaneously. Typically lasting 12-18 months, dual coding ensures continuity for billing, reporting, and research while building ICD-11 competency. It doubles work temporarily but prevents dangerous gaps.

Crosswalk A mapping between ICD-10 and ICD-11 codes. Simple crosswalks show one-to-one correspondences. Complex crosswalks require clinical interpretation for one-to-many or many-to-one relationships. No crosswalk achieves perfect translation—clinical review remains essential.

Case Example 3: Crosswalk Complexity

The billing department discovers their automated crosswalk fails for eating disorders. ICD-10's F50.01 (Anorexia nervosa, restricting type) maps to multiple potential ICD-11 codes:

- 6B80.0 (Anorexia nervosa, restricting pattern)

- 6B80.1 (Anorexia nervosa, binge-purge pattern)
- Severity qualifiers needed
- Remission status options

They implement clinical review for all eating disorder codes, ensuring accurate mapping based on full documentation rather than automated translation.

Go-Live The date an organization begins using ICD-11 for production coding. Different from training starts or testing begins, go-live means real patient records receive ICD-11 codes. Most organizations phase go-live by department or facility rather than enterprise-wide conversion.

Brownout Period The risky phase when some systems use ICD-11 while others expect ICD-10. During brownouts, interfaces fail, reports break, and claims reject. Careful planning minimizes brownout duration and impact.

Super User Staff members with advanced ICD-11 training who support colleagues during implementation. Beyond their regular duties, super users answer questions, demonstrate techniques, and bridge between IT and clinical staff. Successful implementations cultivate robust super user networks.

Clinical Integration Terms

Natural Language Processing (NLP) Computer interpretation of clinical narrative to suggest appropriate codes. ICD-11's structure improves NLP accuracy by providing consistent patterns and rich semantic relationships. Modern NLP can parse complex clinical notes and propose complete code clusters.

Clinical Decision Support (CDS) Automated guidance triggered by ICD-11 codes. When specific diagnoses enter the record, CDS can suggest relevant orders, remind about guidelines, or flag quality measures. ICD-11's precision enables more targeted, less intrusive CDS.

Quality Measures Standardized metrics for healthcare performance. ICD-11's specificity improves quality measurement by better identifying relevant populations. Post-coordination ensures patients with complex conditions get counted correctly for all applicable measures.

Language shapes thought. As you internalize these terms, you'll find yourself thinking differently about clinical documentation. The vocabulary of ICD-11 pushes toward precision, relationships, and semantic meaning. Master these terms not just as definitions but as tools for better clinical communication.

Key Takeaways

- Foundation Component contains complete disease knowledge while linearizations provide practical coding subsets
- Stem codes form the four-character base that can stand alone or accept extensions
- Post-coordination combines codes using forward slash (/) and ampersand (&) symbols
- URIs provide permanent identifiers enabling cross-version and cross-language tracking
- APIs make ICD-11 a dynamic service rather than static publication
- Dual coding periods ensure safe transition while building competency
- Crosswalks require clinical interpretation, not just automated mapping
- Technical terms like FHIR and JSON enable seamless system integration
- Clinical terms like NLP and CDS show how ICD-11 improves care delivery

Chapter 15: Frequently Asked Questions

Years of ICD-11 training sessions reveal the same concerns surfacing repeatedly. Behind polite professional questions lie real anxieties—about job security, competency, and patient care during transition. This chapter addresses not just the surface questions but the deeper worries they represent.

When will ICD-11 become mandatory in the United States?

The short answer: No official mandate exists yet (45). The realistic answer requires understanding how U.S. healthcare adoption works.

Unlike countries with centralized health systems, the U.S. follows a complex dance between government agencies, private payers, and provider organizations. The Centers for Medicare & Medicaid Services (CMS) drives adoption through reimbursement requirements. When CMS requires ICD-11 for payment, the market follows.

Current indicators suggest 2025-2027 for the CMS proposed rule, with implementation 2-3 years later. This timeline allows for:

- Technology vendor preparation
- Workforce training
- Dual coding transitions
- Payer readiness

But don't wait for mandates. Early adopters gain competitive advantages through:

- Better clinical documentation
- Improved research capabilities
- Enhanced quality reporting
- Preparation time without pressure

Case Example 1: Strategic Early Adoption

Mountain Health Network started ICD-11 preparation in 2024, well before mandates. CEO Jennifer Williams explains their reasoning: "We watched organizations scramble with ICD-10. The pain was unnecessary. By starting early, we control our timeline, train thoroughly, and negotiate better vendor terms. When mandates come, we'll be teaching others rather than struggling ourselves" (46).

Their phased approach:

- 2024: Leadership education and vendor assessment
- 2025: Infrastructure upgrades and pilot departments
- 2026: Gradual rollout with dual coding
- 2027: Full implementation before mandates

Early preparation transformed a compliance burden into strategic opportunity.

How do I handle post-coordination?

Post-coordination intimidates experienced coders accustomed to finding pre-combined codes. The fear is understandable—you're moving from selection to construction. But post-coordination follows logical patterns anyone can master.

Start simple. Learn one symbol at a time:

- Master the forward slash (/) for showing relationships
- Then add the ampersand (&) for extensions
- Practice with common scenarios from your specialty
- Build complexity gradually

Think of post-coordination like writing sentences. You learned grammar rules, then practiced until writing became natural. Post-coordination follows the same progression—awkward at first, then automatic.

Common patterns emerge quickly:

- Diabetes with complications: 5A11/[complication codes]
- Injuries with external causes: [injury]/[cause]&[details]
- Infections with organisms: [infection site]/[organism]

Case Example 2: Learning Through Patterns

Coder Sarah Martinez felt overwhelmed by post-coordination until she recognized patterns. She created specialty-specific templates:

For cardiology:

- Acute MI: BA41/[location]&[vessel]
- Heart failure: BD11/[cause]&[stage]
- Arrhythmias: BC81/[triggers]&[symptoms]

For orthopedics:

- Fractures: [bone code]&[laterality]&[healing status]
- Joint disorders: [joint code]/[cause]&[severity]

Templates gave structure while she built confidence. Within weeks, she created novel combinations for complex cases without consulting references.

Tools help during transition:

- Validation software prevents invalid combinations
- Smart interfaces suggest appropriate extensions
- Practice environments allow safe experimentation
- Peer review catches early mistakes

Will productivity drop permanently?

Productivity concerns keep coding managers awake. Studies show initial productivity drops 20-30% (47). But "initial" is the key word. Recovery follows predictable patterns:

Month 1-2: Significant slowdown as coders think through each code
Month 3-4: Gradual improvement as patterns become familiar
Month 5-6: Return to baseline productivity **Month 7+**: Productivity exceeds ICD-10 levels

Why does productivity eventually improve? ICD-11's logical structure eliminates time wasted searching for codes that don't exist. Post-coordination removes hunting through inadequate pre-combined options. Consistency reduces memorization burden.

Case Example 3: Productivity Recovery Curve

Regional Medical Center tracked coder productivity meticulously:

Baseline (ICD-10): 24 charts per day average Month 1 (ICD-11): 16 charts per day (-33%) Month 3: 20 charts per day (-17%) Month 6: 24 charts per day (returned to baseline) Month 9: 28 charts per day (+17% improvement)

Coding Manager David Chen analyzed the improvement: "ICD-11's structure makes sense. Once coders stop translating from ICD-10 thinking, they code naturally. The consistency means less lookup time. Post-coordination eliminates arguing about which inadequate code fits best" (48).

Factors accelerating productivity recovery:

- Robust training before go-live
- Strong super user support
- Graduated complexity during transition
- Investment in productivity tools
- Celebrating early wins

What about small practices with limited resources?

Small practices face unique challenges but also unique advantages. Without large IT departments or dedicated coding staff, ICD-11 seems daunting. Yet small practices often adapt faster than large organizations.

Advantages of being small:

- Fewer systems to update
- Direct communication paths
- Simplified decision-making
- Flexibility in approaches
- Lower complexity cases typically

Strategies for resource-limited settings:

- Partner with similar practices for shared training
- Use vendor-provided education materials
- Leverage free WHO resources
- Implement gradually by provider
- Focus on highest-volume codes first

Cloud-based solutions level the playing field. Small practices access the same coding tools as large systems without infrastructure investment. Subscription models spread costs over time.

How will ICD-11 affect medical billing?

Billing touches everyone's anxiety. Will claims be denied? Will revenue drop? Will payers be ready? Historical experience with ICD-10 provides lessons.

Initial disruption is likely but manageable:

- Some payers will lag in readiness
- Claim edits will need updating
- Denial rates may temporarily increase
- Cash flow requires closer monitoring

Mitigation strategies that work:

- Early payer engagement
- Robust crosswalk maintenance
- Increased claim scrubbing
- Enhanced denial management
- Clear documentation improvement

ICD-11's precision ultimately improves billing by:

- Better supporting medical necessity
- Reducing ambiguous diagnoses
- Improving risk adjustment accuracy
- Enhancing prior authorization success

Do I need to learn everything at once?

Absolutely not. ICD-11 is vast—nobody masters it entirely. Focus on what matters for your role:

For physicians: Learn your specialty's common codes and basic post-coordination **For coders**: Master the structure, then build specialty expertise **For analysts**: Understand mapping impacts on reports **For administrators**: Grasp operational implications **For IT staff**: Focus on technical architecture

Progressive learning works better than comprehensive cramming. Start with frequently used codes. Add complexity as comfort grows. Build on success rather than struggling with everything simultaneously.

Will my current certifications still be valid?

Professional certifications remain valid but require updating. Major certifying bodies announced transition plans:

- AHIMA: Current CCS/CCS-P remain active, add ICD-11 competency
- AAPC: CPC/COC credentials continue, ICD-11 specialization available

- Specialty boards: Gradual integration into existing programs

Maintaining credentials requires:

- Completing ICD-11 education modules
- Passing competency assessments
- Earning continuing education credits
- Demonstrating practical application

View this as career enhancement, not replacement. ICD-11 expertise adds to your qualifications rather than invalidating existing knowledge.

What if our vendors aren't ready?

Vendor readiness varies dramatically. Some lead the market with robust solutions. Others lag dangerously. Protect your organization through:

Clear requirements: Define exactly what ICD-11 capabilities you need **Firm timelines**: Include penalties for delays **Proof of concept**: Require demonstrations, not promises **Reference checks**: Talk to other clients about actual experiences **Contingency plans**: Prepare for vendor failure

Never accept "we'll be ready" without evidence. Request detailed roadmaps. Attend user group meetings. Monitor vendor financial stability—some won't survive the transition.

How do I stay current with updates?

ICD-11 updates annually, unlike ICD-10's sporadic changes. Staying current requires systematic approaches:

- Subscribe to WHO update notifications
- Join professional association committees
- Attend annual update training

- Build internal review processes
- Designate update coordinators

Updates typically include:

- New codes for emerging conditions
- Refinements to existing codes
- Additional extension options
- Clarified coding guidelines
- Enhanced semantic relationships

Create organizational memory for updates. Document why changes matter for your setting. Train incrementally rather than annually.

What's the single most important thing to know?

If forced to choose one crucial insight: ICD-11 is about relationships, not just codes. Understanding how conditions connect—through causation, manifestation, timing—matters more than memorizing individual codes. Once you grasp the relationship model, everything else follows logically.

This shift from lists to relationships transforms clinical documentation. You stop forcing complex reality into simple boxes. Instead, you build accurate representations of what you actually see in patients.

Questions reflect concerns, and concerns deserve honest answers. ICD-11 brings challenges—denying them helps nobody. But challenges accompany opportunities. Organizations and individuals who engage thoughtfully with these questions position themselves for success. Your concerns are valid, your questions important, and your success achievable.

Key Takeaways

- U.S. mandate timeline points to 2025-2027 for rules, implementation 2-3 years later

- Post-coordination follows learnable patterns—start simple and build complexity
- Productivity drops initially but exceeds ICD-10 levels within 6-9 months
- Small practices can succeed through cloud solutions and collaborative approaches
- Billing disruption is temporary; precision ultimately improves reimbursement
- Progressive learning by role works better than trying to master everything
- Current certifications remain valid but require ICD-11 competency additions
- Vendor readiness varies—require proof, not promises
- Annual updates need systematic organizational processes
- Understanding relationships matters more than memorizing codes

Appendix A: Complete ICD-10 to ICD-11 Mapping Tables

Traditional mapping tables fill hundreds of pages with cryptic correspondences. This appendix takes a different approach— explaining the logic behind mappings so you can predict conversions without constant lookups. Understanding why codes map as they do proves more useful than memorizing thousands of individual translations.

Understanding Mapping Complexity

Mappings between ICD-10 and ICD-11 fall into four categories, each requiring different approaches. Simple one-to-one mappings seem straightforward but hide subtle differences. One-to-many mappings force choices based on clinical detail. Many-to-one mappings lose granularity unless rescued by post-coordination. Complex mappings require complete rethinking.

The most challenging aspect isn't finding corresponding codes—it's preserving clinical meaning. ICD-10's pre-coordinated approach attempted to capture complex scenarios in single codes. ICD-11's post-coordination philosophy builds complex scenarios from simpler elements. This fundamental difference drives mapping complexity.

Case Example 1: Diabetes Mapping Evolution

Consider how diabetes with complications maps between systems. ICD-10's E11.21 combines three concepts: Type 2 diabetes + complication + specific complication (diabetic nephropathy). This single code becomes multiple elements in ICD-11:

The base condition: **5A11** (Type 2 diabetes mellitus) The complication relationship: / (forward slash indicating causation) The specific complication: **5A25** (Diabetic kidney disease) Additional detail: Stages, control status, treatment modifiers

Dr. Patricia Kim, endocrinologist, explains the clinical impact: "ICD-10 forced us to choose one primary complication when patients had several. Now I can document complete reality—diabetes causing retinopathy, neuropathy, and nephropathy, each with its own severity. The mapping looks complex, but it enables truth" (49).

Primary Care Common Conversions

Primary care encounters high-volume, bread-and-butter diagnoses. These conversions affect every practice:

Respiratory conditions show systematic reorganization. Upper respiratory infections, previously scattered across J codes, consolidate logically. J00 (Acute nasopharyngitis) maps to CA01, but ICD-11 adds temporal pattern options—first episode, recurrent, or persistent. J06.9 (Acute upper respiratory infection, unspecified) pushes toward specificity with CA0Z, requiring at least anatomical location.

Hypertension gains nuance. I10 (Essential hypertension) seems to map simply to BA00, but ICD-11 expects blood pressure documentation. The code alone no longer suffices—you indicate control status, treatment response, and target organ involvement through extensions.

Mental health undergoes philosophical shifts. Depression illustrates this transformation. F32.0 (Mild depressive episode) becomes 6A70.0 (Single episode depressive disorder, mild), but ICD-11 adds dimensional assessments—mood, physical, cognitive symptoms separately rated. Anxiety disorders similarly expand from categories to continua.

Case Example 2: Pediatric Practice Conversion

Dr. Sarah Johnson converted her pediatric practice's twenty most common diagnoses. Her experience revealed patterns:

"Acute otitis media looked simple—H66.90 to AA20. But ICD-11 wanted laterality, causative organism when known, and treatment response. Initially frustrating, but it pushed us toward better documentation. Now we track which kids get recurrent bacterial versus viral ear infections. Treatment patterns improved" (50).

Her conversion revealed:

- Childhood infections gain temporal patterns
- Developmental conditions require functional assessments
- Injury codes demand external cause integration
- Preventive codes expand dramatically

Chronic Disease Mapping Patterns

Chronic diseases reveal ICD-11's relationship focus. Single ICD-10 codes explode into networks of connected conditions.

Cardiovascular diseases exemplify this expansion. I25.10 (Atherosclerotic heart disease of native coronary artery without angina pectoris) maps to BA80.Z, but ICD-11 expects:

- Specific vessels involved
- Degree of stenosis
- Functional impact
- Relationship to other conditions

The mapping forces cardiovascular documentation improvement. Cardiologists must specify which vessels, what percentage blocked, whether causing symptoms, and connections to hypertension, diabetes, or lipid disorders.

Chronic kidney disease moves from stages to comprehensive assessment. N18.3 (Chronic kidney disease, stage 3) translates to GB61.3, but adds:

- Underlying cause specification
- Albuminuria levels
- Progression rate
- Complications present

Chronic pain finally receives proper classification. ICD-10 scattered pain codes across chapters. ICD-11 creates coherent organization:

- MG30 series for acute pain by location
- MG70 series for chronic primary pain
- Secondary pain linked to underlying conditions
- Functional impact assessments

Specialty-Specific Mapping Challenges

Each specialty faces unique mapping complexities reflecting their domain evolution.

Oncology deals with precision medicine integration. ICD-10's morphology codes become integral to ICD-11 classification. C50.911 (Malignant neoplasm of unspecified site of right female breast) requires:

- Histological type specification
- Molecular markers
- Stage integration
- Treatment phase indicators

Oncologists report initial frustration followed by appreciation. The detailed mapping enables research enrollment, treatment selection, and outcome tracking previously requiring separate registries.

Case Example 3: Surgical Practice Mapping

Dr. Robert Chen's surgical group discovered procedure-diagnosis relationships needed complete reconceptualization. "ICD-10 let us code 'status post' conditions vaguely. ICD-11 demands precision—what surgery, when, what complications, current status. Our breast cancer post-mastectomy patients now have codes reflecting reconstruction type, cosmetic outcomes, lymphedema presence. It tells their whole story" (51).

Neurology benefits from ICD-11's symptom-disease relationships. Seizure disorders illustrate the improvement:

- G40.909 (Epilepsy, unspecified, not intractable, without status epilepticus) fragments into specific seizure types
- Each type links to underlying causes when known
- Seizure frequency and impact on daily life join the code cluster
- Medication responsiveness becomes documentable

Psychiatry undergoes the most dramatic changes. DSM-5 alignment brings dimensional assessment to formerly categorical conditions. Mapping requires understanding conceptual shifts, not just code changes. Personality disorders, eating disorders, and substance use disorders all require new thinking about severity, course, and functional impact.

Migration Strategy Lessons

Successful mapping requires more than code tables. Organizations that navigate smoothly share common strategies:

Clinical involvement throughout: Pure administrative mapping fails. Clinicians must validate that mapped codes preserve clinical meaning. A diabetes nurse educator catches nuances that coders miss.

Specialty-specific validation: Each department reviews their high-volume diagnoses. Patterns emerge that inform training and documentation improvement.

Iterative refinement: Initial mappings rarely achieve perfection. Plan review cycles, gathering feedback and adjusting mappings based on real-world use.

Documentation improvement parallel track: Many ICD-10 codes map poorly because original documentation lacked detail. Use mapping as opportunity to enhance clinical documentation.

Acceptance of imperfection: Some clinical concepts simply don't translate cleanly. Document these situations and develop consistent organizational approaches rather than seeking impossible perfect mappings.

Think of mapping not as mechanical translation but as clinical evolution. Each mapping decision reflects advancing medical understanding. The complexity serves a purpose—capturing clinical reality more accurately. When mapping frustrates, ask what clinical truth ICD-11 seeks that ICD-10 obscured. That insight transforms tedious conversion into meaningful improvement.

Key Takeaways

- Mapping complexity stems from fundamental philosophy differences between ICD-10's pre-coordination and ICD-11's post-coordination
- One-to-many mappings require clinical judgment to select appropriate detail level
- Primary care conversions push toward greater specificity in common conditions
- Chronic disease mappings reveal ICD-11's relationship-focused approach
- Specialty mappings reflect domain-specific advances in medical understanding
- Successful migration requires clinical validation, not just administrative conversion
- Documentation improvement must parallel mapping efforts
- Perfect mapping rarely exists—consistency matters more than perfection
- Understanding mapping logic enables prediction without memorization

Appendix B: Implementation Checklist

Implementation succeeds through systematic progression, not heroic efforts. This checklist, refined through dozens of implementations, prevents common oversights while maintaining momentum. Each item represents lessons learned—often painfully—by organizations that went before.

24 Months Before Go-Live

The foundation phase seems premature to those eager for action. Resist that urgency. These early decisions cascade through entire implementation.

Executive Alignment

- Secure C-suite sponsor who understands ICD-11 as business transformation, not IT project
- Establish board-level reporting expectations
- Define success metrics beyond technical implementation
- Allocate realistic budget including productivity impact
- Communicate strategic importance across leadership

Stakeholder Identification

- Map every department using diagnosis codes
- Identify external partners requiring coordination
- Document current workflows touching ICD codes
- Assess change readiness by stakeholder group
- Create comprehensive communication matrix

Current State Assessment

- Inventory all systems containing ICD codes
- Document existing coding productivity benchmarks
- Analyze diagnosis documentation quality
- Review payer contract language

- Evaluate current training resources

Case Example 1: Foundation Phase Discovery

Metro Health thought they understood their ICD footprint until systematic assessment revealed surprises. Quality Director Lisa Anderson recalls: "We found a researcher running a cardiac database since 2015, completely outside our main systems. Marketing was analyzing diagnosis patterns for service line planning. Even our cafeteria used ICD codes for special diets. Missing any of these would have caused failures" (52).

Their assessment uncovered:

- 47 systems using ICD codes (expected 15)
- 12 external data feeds requiring coordination
- 300+ reports dependent on diagnosis coding
- 23 payer contracts referencing ICD specifically

Early discovery enabled proper planning rather than crisis management.

18 Months Before Go-Live

With foundations established, preparation intensifies. This phase separates smooth implementations from chaotic scrambles.

Governance Structure Launch

- Charter executive steering committee
- Appoint dedicated project manager
- Establish clinical advisory groups by specialty
- Create technical working teams
- Define decision-making processes

Vendor Engagement

- Request detailed ICD-11 roadmaps from all vendors

- Require proof-of-concept demonstrations
- Negotiate implementation support agreements
- Establish vendor accountability metrics
- Plan for vendor failure contingencies

Resource Planning

- Finalize implementation budget
- Identify dedicated team members
- Plan productivity support coverage
- Allocate training time by role
- Secure technical infrastructure funding

Communication Launch

- Begin awareness campaigns
- Share implementation timeline broadly
- Establish feedback channels
- Create ICD-11 resource center
- Celebrate early milestones

12 Months Before Go-Live

The building phase transforms plans into reality. Momentum accelerates, requiring careful coordination.

Technical Infrastructure

- Upgrade servers to meet processing requirements
- Expand storage for dual coding period
- Enhance network capacity for API traffic
- Implement test environments
- Establish backup and recovery procedures

Mapping Development

- Create specialty-specific mapping tables
- Validate high-volume diagnosis conversions

- Document complex mapping decisions
- Establish mapping governance process
- Plan crosswalk maintenance procedures

Training Design

- Develop role-specific curricula
- Create hands-on practice materials
- Build competency assessment tools
- Identify super users by department
- Design ongoing education programs

Case Example 2: Building Phase Coordination

Regional Medical Network learned coordination criticality when their building phase nearly derailed. Project Manager David Kim explains: "Three teams were building solutions independently—IT creating interfaces, Education developing training, and Operations designing workflows. When we integrated, nothing aligned. We lost two months rebuilding with daily coordination meetings" (53).

Recovery required:

- Daily stand-ups across all teams
- Integrated project planning tools
- Shared testing scenarios
- Joint design sessions
- Unified change control

6 Months Before Go-Live

Preparation transitions to execution. The pace intensifies as go-live approaches.

Pilot Planning

- Select pilot departments
- Define pilot success criteria

- Create pilot measurement tools
- Establish feedback mechanisms
- Plan pilot expansion strategy

Training Launch

- Begin super user training
- Conduct leadership orientation
- Start end-user education waves
- Implement competency testing
- Track training completion rates

Technical Testing

- Complete unit testing
- Perform integration testing
- Conduct performance testing
- Execute user acceptance testing
- Stress test infrastructure

Dual Coding Preparation

- Finalize dual coding workflows
- Test crosswalk automation
- Train dual coding teams
- Establish quality monitoring
- Plan productivity support

3 Months Before Go-Live

Final preparations demand flawless execution. Details determine success.

Pilot Execution

- Launch department pilots
- Monitor pilot metrics daily
- Gather continuous feedback

- Refine based on lessons
- Prepare broader rollout

Communication Intensification

- Increase update frequency
- Share pilot successes
- Address concerns transparently
- Countdown communications
- Build excitement momentum

Go-Live Readiness

- Finalize cutover plans
- Assign go-live support teams
- Create issue tracking systems
- Establish command center
- Plan celebration events

Case Example 3: Final Phase Excellence

Children's Hospital created a go-live command center that became the implementation model. Operations Director Maria Santos describes: "We converted our boardroom into mission control. Real-time dashboards, dedicated phone lines, clinical experts on-site, pizza delivered regularly. When issues arose—and they did—we resolved them in minutes, not days" (54).

Command center elements:

- Executive presence for rapid decisions
- Technical team for immediate fixes
- Clinical experts for coding questions
- Communication team for updates
- Celebration bell for successes

Go-Live Execution

The culmination requires precision while maintaining flexibility for inevitable surprises.

Day 1 Critical Tasks

- Command center activation at least 2 hours before first shift
- System health verification
- Support team deployment
- Issue tracking initialization
- Success metric monitoring

Week 1 Priorities

- Daily leadership rounds
- Productivity monitoring
- Issue pattern analysis
- Quick win celebrations
- Fatigue management

Month 1 Stabilization

- Workflow optimization
- Training reinforcement
- Productivity recovery support
- Quality audits
- Stakeholder satisfaction assessment

Post-Implementation Sustainability

Success extends beyond go-live. Long-term excellence requires continuous attention.

90-Day Optimization

- Comprehensive lessons learned
- Workflow refinements
- Advanced feature activation
- Productivity analysis

- ROI documentation

Annual Maintenance

- Update process establishment
- Continuing education programs
- Quality monitoring systems
- Vendor relationship management
- Strategic advancement planning

This checklist represents collective wisdom from numerous implementations. Every item prevents specific problems encountered by others. Use it as foundation, but customize for your unique environment. Mark progress visibly—completed checkmarks build momentum and confidence. Most importantly, start now. Time enables thoughtful implementation; urgency forces costly shortcuts.

Key Takeaways

- Begin foundation work 24 months before go-live to prevent rushed decisions
- Early stakeholder identification reveals hidden complexities
- Vendor readiness requires verification through demonstrations, not promises
- Technical infrastructure needs often exceed initial estimates
- Pilot implementations reveal issues safely before organization-wide impact
- Command center approach enables rapid issue resolution during go-live
- Post-implementation optimization continues indefinitely
- Each checklist item prevents specific problems learned through experience
- Customization for organizational uniqueness improves standard checklist

Appendix C: Sample Training Materials

Effective training transforms anxiety into capability. These sample materials, refined through hundreds of training sessions, accelerate learning while building confidence. Adapt them to your organization's culture, but maintain the proven pedagogical principles they embody.

Core Curriculum Architecture

Adult learners need relevance, not theory. Structure training around their actual work, not abstract concepts. Each module follows a consistent pattern: why this matters, what you need to know, how to apply it, and practice until comfortable.

Module 1: Why ICD-11 Exists (2 hours)

Learning objectives that resonate:

- Explain how ICD-11 improves your specific clinical documentation
- Identify three benefits for your department
- Recognize the business case for change

Skip the history lesson. Focus on "what's in it for me?" Frame ICD-11 as solving their current frustrations—inadequate codes, complex workarounds, documentation queries. Use their actual problem scenarios to demonstrate improvements.

Case Example 1: Specialty-Specific Relevance

Emergency physician Dr. James Chen created training that clicked: "I showed our ED docs their most annoying ICD-10 scenarios—chest pain with unclear etiology, multiple trauma inadequately captured, vague abdominal pain. Then demonstrated ICD-11 solutions. Engagement soared when they saw personal benefit" (55).

His approach:

- Collected frustrating coding scenarios from each physician
- Demonstrated ICD-11 solutions for their specific cases
- Let them practice on their own challenging patients
- Celebrated when they found better codes

Module 2: Code Structure Mastery (3 hours)

Foundation knowledge delivered practically:

- Decode any ICD-11 code in 30 seconds
- Build basic post-coordination
- Recognize valid versus invalid combinations

Avoid memorization. Teach patterns. Like learning a language, grammar matters more than vocabulary. Once learners grasp structure, they can construct any code they need.

Interactive exercises that work:

1. Code structure puzzles—identify the chapter, block, and category
2. Build-a-code workshops—construct progressively complex codes
3. Error detection games—find the invalid combinations
4. Speed challenges—decode codes against the clock

Module 3: Post-coordination Power (4 hours)

The most feared topic becomes the most appreciated when taught properly:

- Start with single extensions
- Progress to dual relationships
- Build complex clusters
- Validate with real cases

Use visual demonstrations. Show post-coordination as building blocks, not complex formulas. Physical manipulatives—actual blocks or cards—help kinesthetic learners grasp abstract concepts.

Module 4: Workflow Integration (3 hours)

Knowledge without application frustrates everyone:

- Map current workflows
- Identify ICD-11 touch points
- Practice integrated scenarios
- Optimize for efficiency

Case Example 2: Workflow-Based Learning

Coding manager Patricia Williams revolutionized training by embedding it in actual workflows: "We recorded coders working through real charts in ICD-10. Then showed the same charts coded in ICD-11. Seeing their own workflows improved convinced skeptics. Training completion jumped from 60% to 95%" (56).

Elements that drove success:

- Screen recordings of actual work
- Side-by-side workflow comparisons
- Time studies showing efficiency gains
- Peer testimonials from early adopters

Assessment Tools That Actually Assess

Traditional multiple-choice tests measure memorization, not capability. Effective assessment mirrors real work.

Knowledge Checks That Matter

Instead of: "Which character indicates chapter in ICD-11?" Ask: "Your patient has diabetes with eye problems. Build the appropriate code."

Instead of: "Define post-coordination." Ask: "Code this scenario: Heart attack caused by cocaine use, affecting the anterior wall."

Practical Competency Demonstrations

Create assessment scenarios from actual cases:

- Common diagnoses from their specialty
- Complex cases requiring post-coordination
- Time-limited coding challenges
- Error correction exercises

Set realistic standards:

- 80% accuracy on common codes
- 70% accuracy on complex scenarios
- Improvement trend more important than perfection
- Focus on safe practice, not speed initially

Case Example 3: Competency Through Cases

Dr. Michael Rodriguez developed case-based assessments that predicted real-world success: "Multiple choice told us nothing. We created 20 cases from each specialty—10 routine, 5 moderate complexity, 5 challenging. Learners coding at 75% accuracy performed well post-implementation. Those below 60% needed support" (57).

Assessment insights:

- Case complexity predicts support needs
- Timed assessments reveal workflow readiness
- Error patterns guide remediation
- Peer review enhances learning

Role-Specific Learning Paths

One-size training fits no one. Tailor content to roles while maintaining consistency.

Physician Path (8 hours total)

- Module 1: Clinical documentation improvements (2 hours)
- Module 2: Common codes for your specialty (3 hours)
- Module 3: Post-coordination basics (2 hours)
- Module 4: Efficiency techniques (1 hour)

Focus on clinical benefits. Minimize administrative burden. Emphasize how better coding improves patient care and reduces queries.

Professional Coder Path (40 hours total)

- Modules 1-4: Complete foundation (12 hours)
- Module 5: Advanced post-coordination (6 hours)
- Module 6: Specialty deep dives (8 hours)
- Module 7: Quality and compliance (6 hours)
- Module 8: Productivity optimization (8 hours)

Build expertise systematically. Balance accuracy with productivity. Create coding artists, not just technicians.

Analyst Path (16 hours total)

- Module 1: ICD-11 data structure (4 hours)
- Module 2: Mapping impacts (4 hours)
- Module 3: Report modifications (4 hours)
- Module 4: Quality metrics (4 hours)

Emphasize data continuity. Show how ICD-11 enhances analytics. Prepare for transition period challenges.

Delivery Modalities That Engage

Mix methods to maintain energy and accommodate learning styles.

172

Self-Paced e-Learning

- Foundation concepts
- Individual practice
- Knowledge checks
- Resource libraries

Best for: Consistent content delivery, flexible scheduling, baseline knowledge

Virtual Instructor-Led

- Complex topics
- Q&A sessions
- Group exercises
- Peer interaction

Best for: Geographically dispersed teams, focused topics, cost efficiency

In-Person Workshops

- Hands-on practice
- Workflow integration
- Team building
- Competency validation

Best for: Skill building, change management, department cohesion

At-the-Elbow Support

- Real-time guidance
- Workflow optimization
- Confidence building
- Issue resolution

Best for: Go-live support, struggling learners, workflow refinement

Making Training Stick

Learning happens after class ends. Build reinforcement into daily work.

Quick Reference Tools

- Laminated code structure guides
- Specialty-specific common codes
- Post-coordination decision trees
- Error prevention checklists

Peer Learning Networks

- Department champions
- Coding circles
- Question channels
- Success sharing

Continuous Reinforcement

- Weekly tips
- Monthly challenges
- Quarterly updates
- Annual recertification

These materials succeed because they respect adult learners. They connect to real work, solve actual problems, and build practical skills. Customize fearlessly—your organization's culture and needs are unique. But maintain the core principle: make learning immediately applicable to improve both confidence and competence.

Key Takeaways

- Structure training around actual workflows, not abstract concepts

- Adult learners need relevance—show "what's in it for me" immediately
- Case-based assessments predict real-world performance better than tests
- Role-specific paths respect different needs while maintaining standards
- Mixed delivery modalities accommodate learning styles and schedules
- Post-coordination becomes manageable when taught as building blocks
- Peer learning networks sustain knowledge after formal training ends
- Quick reference tools bridge the gap between learning and doing
- Continuous reinforcement prevents skill decay

Appendix D: Vendor Resource Directory

Selecting the right vendors determines implementation success as much as internal preparation. This directory, compiled from extensive market analysis and user experiences, guides selection while avoiding costly mistakes. Vendors evolve rapidly—verify current capabilities rather than relying on historical reputation.

EHR Vendor ICD-11 Readiness

Electronic health record vendors anchor your implementation. Their readiness—or lack thereof—shapes your entire strategy.

Major EHR Players

Epic Systems leads market readiness with native ICD-11 functionality built into recent versions. Their approach integrates post-coordination naturally into clinical workflows. Users report minimal disruption during transition. However, older Epic installations require significant upgrades. Budget 6-12 months for Epic's implementation timeline.

Oracle Health (formerly Cerner) takes a modular approach, allowing staged implementation. Their ICD-11 solution emphasizes backward compatibility, easing transition anxiety. Post-coordination tools need refinement, according to early adopters. Smaller facilities find Oracle's solution more flexible than Epic's comprehensive overhaul.

Athenahealth leverages cloud architecture for rapid ICD-11 deployment. Updates happen seamlessly without on-site installation. Their smaller market share means fewer peer organizations to learn from. Best suited for ambulatory settings prioritizing simplicity over customization.

NextGen Healthcare targets ambulatory practices with pragmatic ICD-11 tools. Their solution emphasizes common codes over comprehensive capability. Specialty practices may find limitations. Price point attracts smaller organizations.

Case Example 1: EHR Selection Impact

Mountain View Medical Group's EHR decision shaped their entire implementation. CIO Jennifer Thompson explains: "We nearly chose based on price alone. Thankfully, we visited three practices using each vendor. Seeing actual workflows revealed huge differences. The cheapest option would have cost us millions in workarounds" (58).

Their evaluation revealed:

- Vendor A: Great demo, poor real-world post-coordination
- Vendor B: Expensive but seamless clinical integration
- Vendor C: Required extensive customization
- Final choice: Vendor B saved money long-term through efficiency

Specialized Coding Software

Professional coding software bridges gaps in EHR functionality while optimizing coder productivity.

3M Health Information Systems dominates enterprise coding software. Their CodeFinder suite handles complex post-coordination elegantly. Natural language processing suggests appropriate codes from clinical text. Investment is substantial but returns through productivity gains. Large health systems report 20-30% efficiency improvements.

Optum360 competes through comprehensive content and analytics. Their Computer-Assisted Coding (CAC) platform learns from your coding patterns. Integration with revenue cycle tools streamlines billing. Mid-size organizations appreciate the balance of capability and cost.

Dolbey focuses on physician documentation improvement alongside coding. Their fusion of speech recognition with ICD-11 coding assists at the point of documentation. Smaller footprint suits targeted implementations.

TruCode serves smaller organizations with subscription-based encoding. Cloud delivery eliminates infrastructure requirements. Limited customization frustrates complex organizations. Perfect for practices seeking simplicity.

Training Organizations

Quality training accelerates adoption while preventing costly mistakes. These organizations proved their ICD-11 expertise through successful implementations.

AHIMA (American Health Information Management Association) sets the standard for coding education. Their ICD-11 certification programs validate competency. Corporate training packages scale from small practices to health systems. Pricing reflects premium positioning but delivers consistent quality.

AAPC (American Academy of Professional Coders) emphasizes practical application over theory. Their boot camps create confident coders quickly. Online options accommodate distributed workforces. Members receive ongoing support through local chapters.

Case Example 2: Training Investment Returns

Regional Health Network compared training approaches across their facilities. Training Director Robert Lee discovered: "Facilities using premium training programs achieved proficiency 40% faster. The $50,000 'saved' by choosing cheaper training cost us $400,000 in extended productivity loss" (59).

Training quality indicators:

- Instructor credentials and experience

178

- Hands-on practice opportunities
- Role-specific customization
- Post-training support included
- Measurable competency validation

HCPro delivers focused training for specific roles. Their materials excel at translating complex concepts simply. Subscription models provide ongoing education. Best for organizations wanting continuous learning versus one-time training.

Local Universities increasingly offer ICD-11 programs through health information management departments. Academic rigor ensures comprehensive coverage. Scheduling accommodates working professionals. Costs typically lower than commercial training.

Consulting Firms

When internal expertise falls short, consultants fill gaps. Choose based on specific needs rather than general reputation.

Leading Healthcare Consultancies

Deloitte brings enterprise-scale implementation expertise. Their methodology, refined through global implementations, reduces risk. Teams include clinical and technical experts. Premium pricing suits complex organizations. Smaller facilities may feel overwhelmed by their approach.

PwC (PricewaterhouseCoopers) emphasizes strategic transformation beyond technical implementation. They excel at change management and stakeholder engagement. International experience benefits multi-national organizations. Domestic-only health systems may find their approach excessive.

Huron Consulting specializes in healthcare, understanding clinical workflows intimately. Their consultants often have provider backgrounds. Right-sized solutions avoid over-engineering. Performance improvement focus delivers measurable ROI.

Case Example 3: Consultant Selection Criteria

Community Health System's failed first attempt taught valuable lessons. CEO Michael Davis reflects: "We hired the biggest name without defining our needs. They delivered a Fortune 500 solution to our community hospital. Second time, we chose consultants who understood our reality. Night and day difference" (60).

Selection criteria that matter:

- Similar client references
- Team members assigned (not just firm reputation)
- Cultural fit with your organization
- Specific ICD-11 experience
- Post-implementation support model

Technology Solution Providers

Beyond core systems, specialized technology accelerates specific aspects of implementation.

Interface Engines

- Mirth Connect: Open-source flexibility for custom integrations
- Rhapsody: Enterprise-scale reliability with ICD-11 optimizations
- Cloverleaf: Traditional solution with proven healthcare record

Analytics Platforms

- Tableau: Visualizes ICD-11 transition impacts elegantly
- PowerBI: Integrates naturally with Microsoft environments
- Qlik: Handles complex ICD-11 relationships well

Mapping Tools

- IMO Health: Clinician-friendly interface mapping

- Intelligent Medical Objects: Comprehensive terminology management
- HDD Access: Specialized ICD-11 mapping validation

Making Vendor Decisions

Vendor selection determines implementation trajectory. Invest time in thorough evaluation:

1. **Define requirements specifically**—generic RFPs yield generic responses
2. **Demand demonstrations with your scenarios**—not vendor-chosen examples
3. **Check references religiously**—especially failed implementations
4. **Negotiate implementation support**—not just software licenses
5. **Plan for vendor failure**—contingencies prevent disasters

No single vendor solves every ICD-11 challenge. Build an ecosystem matching your needs, culture, and resources. The "best" vendor for another organization may fail in your environment. Focus on fit over features, support over sales promises, and proven experience over potential capability.

Key Takeaways

- EHR vendor readiness varies dramatically—verify current capabilities, not promises
- Epic leads comprehensively while smaller vendors offer targeted solutions
- Professional coding software significantly improves productivity when properly selected
- Training quality directly correlates with implementation success speed
- Consulting firms must match organizational size and culture for effectiveness

- Specialized technology tools address specific implementation challenges
- Vendor ecosystems outperform single-vendor solutions
- Reference checking reveals more than demonstrations
- Implementation support matters more than software features
- Plan for vendor failure to prevent implementation disruption

References

1. World Health Organization. (2019). ICD-11 for Mortality and Morbidity Statistics. Geneva: WHO Press.
2. Harrison, J. E., Weber, S., Jakob, R., & Chute, C. G. (2021). ICD-11: an international classification of diseases for the twenty-first century. BMC Medical Informatics and Decision Making, 21(Suppl 6), 206.
3. Fung, K. W., Xu, J., & Bodenreider, O. (2020). The new International Classification of Diseases 11th edition: a comparative analysis with ICD-10 and ICD-10-CM. Journal of the American Medical Informatics Association, 27(5), 738-746.
4. Densen, P. (2011). Challenges and opportunities facing medical education. Transactions of the American Clinical and Climatological Association, 122, 48-58.
5. Chen, J. (2023). Personal communication regarding ICD-11 clinical implementation. Regional Medical Center Quality Report.
6. Nordic Health Data Collaborative. (2023). ICD-11 Implementation Outcomes in Scandinavian Health Systems. Copenhagen: NHDC Publishing.
7. Smith, A. B., & Johnson, C. D. (2022). Workforce impacts of ICD-11 transition: A multi-site study. Health Information Management Journal, 51(3), 145-156.
8. World Health Organization. (2007). ICD-11 Revision Project Plan. Geneva: WHO Document Production Services.
9. Jakob, R. (2018). The development of ICD-11: Historical perspective and innovation. WHO Bulletin, 96(4), 234-242.
10. Tudorache, T., Nyulas, C., Noy, N. F., & Musen, M. A. (2013). WebProtégé: A collaborative ontology editor and knowledge acquisition tool for the web. Semantic Web, 4(1), 89-99.
11. Tanaka, K. (2019). Cultural adaptation in medical classification: Lessons from ICD-11 field trials. International Journal of Medical Informatics, 128, 45-52.
12. Patient Advocacy Coalition. (2018). Patient Voice in ICD-11: A Report on Participation and Outcomes. Brussels: PAC Press.

13. Chute, C. G. (2019). The copernican view of health data: ICD-11 and the digital revolution. Presentation at AMIA Annual Symposium, Washington, DC.
14. Martinez, S. (2022). Impact of ICD-11 immune system chapter on clinical practice. Journal of Clinical Immunology, 42(4), 567-578.
15. Thompson, M. (2020). Sexual health classification in ICD-11: Progress and challenges. International Journal of Sexual Health, 32(3), 234-245.
16. Park, J. (2021). Understanding ICD-11 architecture: Foundation and linearizations. Health Information Science and Systems, 9(1), 18.
17. Wei, C. (2023). Real-world API implementation for ICD-11. Healthcare IT News, 15(3), 22-28.
18. Kim, P. (2023). Post-coordination in oncology: Clinical applications of ICD-11. Journal of Oncology Practice, 19(2), 89-97.
19. Chen, R. (2022). Precision medicine and ICD-11: Aligning classification with targeted therapy. Nature Reviews Clinical Oncology, 19(8), 456-465.
20. World Health Organization. (2020). COVID-19 coding in ICD-11: Emergency use guidance. Geneva: WHO Technical Report.
21. Wu, J. (2023). Digital transformation through ICD-11 APIs. Healthcare Information Management Systems Society Proceedings, 2023, 145-156.
22. Mendoza, C. (2021). Multilingual medical classification: The ICD-11 approach. International Journal of Medical Terminology, 15(4), 234-245.
23. Tanaka, Y. (2023). Rare disease visibility through ICD-11 digital architecture. Orphanet Journal of Rare Diseases, 18(1), 67.
24. Anderson, P. (2022). ICD-10 to ICD-11 mapping: Validation study results. Medical Coding Review, 34(5), 23-31.
25. Chen, J. (2023). Diabetes coding evolution: Clinical implications of ICD-11 transitions. Diabetes Care, 46(3), 456-465.

26. American Psychiatric Association. (2022). ICD-11 mental health classifications: Implementation guide. Washington, DC: APA Publishing.
27. Wilson, R. (2023). Large health system ICD-11 transition: Lessons learned. Journal of Healthcare Information Management, 37(2), 45-56.
28. Thompson, L. (2022). Hidden complexities in ICD-11 implementation. Healthcare Executive, 37(4), 22-30.
29. Chen, M. (2023). Financial planning for ICD-11: A CFO perspective. Healthcare Financial Management, 77(3), 34-42.
30. Williams, P. (2022). Clinical engagement strategies for ICD-11 adoption. Physician Leadership Journal, 9(4), 56-65.
31. Chen, R. (2023). Validation through real cases: ICD-11 pilot outcomes. Quality Management in Healthcare, 32(1), 23-34.
32. Tanasri, S. (2022). Thailand's rapid ICD-11 implementation: Success factors. Asian Pacific Journal of Health Management, 17(2), 45-56.
33. Martinez, J. (2023). Dual coding period management: Technology solutions. Journal of Health Information Technology, 15(2), 78-89.
34. Rodriguez, M. (2023). Cardiology applications of ICD-11 post-coordination. Journal of the American College of Cardiology, 81(5), 456-467.
35. Kim, S. (2023). Precision oncology documentation with ICD-11. Clinical Cancer Research, 29(4), 678-689.
36. Chang, J. (2022). Emergency department ICD-11 implementation. Annals of Emergency Medicine, 79(3), 234-245.
37. Anderson, L. (2023). Complex case management with ICD-11. Case Management Monthly, 20(3), 12-18.
38. Martinez, R. (2023). Mental health coding revolution: ICD-11 in practice. Psychiatric Services, 74(2), 156-165.
39. Anderson, T. (2022). Recovering from failed ICD-11 implementation. Healthcare IT Management, 17(4), 34-42.
40. Wong, P. (2023). Beyond technical training: Conceptual understanding of ICD-11. Medical Informatics Education, 11(2), 89-98.

41. Liu, K. (2023). Infrastructure requirements for ICD-11: A CTO perspective. Healthcare Technology Review, 28(3), 45-56.
42. Peterson, L. (2022). Leveraging WHO Academy for organizational training. Health Education Quarterly, 49(4), 234-245.
43. Martinez, J. (2023). Career advancement through ICD-11 certification. HIM Professional Quarterly, 15(2), 22-28.
44. Chen, R. (2023). Adult learning principles in ICD-11 training. Medical Education Today, 45(3), 156-167.

45. Centers for Medicare & Medicaid Services. (2024). ICD-11 Implementation Planning Update. CMS Quarterly Report, January 2024.
46. Williams, J. (2024). Strategic advantages of early ICD-11 adoption. Healthcare Executive Quarterly, 39(1), 23-31.
47. National Health Information Management Association. (2023). ICD-11 Productivity Impact Study: Multi-site Analysis. NHIMA Research Series, 2023(4).
48. Chen, D. (2024). Productivity recovery patterns in ICD-11 implementation. Journal of Health Information Management, 38(2), 145-156.
49. Kim, P. (2023). Clinical meaning preservation in ICD-10 to ICD-11 mapping. Endocrinology Coding Quarterly, 12(4), 234-242.
50. Johnson, S. (2024). Pediatric practice ICD-11 conversion: Lessons learned. Pediatric Practice Management, 29(3), 167-175.
51. Chen, R. (2023). Surgical diagnosis precision with ICD-11. American Journal of Surgery Informatics, 15(2), 89-97.
52. Anderson, L. (2023). Comprehensive assessment reveals hidden complexity. Quality Management in Healthcare, 32(4), 201-209.
53. Kim, D. (2024). Coordination criticality in ICD-11 building phase. Project Management in Healthcare, 18(1), 45-53.
54. Santos, M. (2023). Command center approach to ICD-11 go-live. Healthcare Operations Excellence, 11(3), 178-186.

55. Chen, J. (2024). Specialty-specific relevance in ICD-11 training. Emergency Medicine Education Journal, 7(2), 123-131.
56. Williams, P. (2023). Workflow-based learning for ICD-11 adoption. Coding Education Quarterly, 15(4), 234-241.
57. Rodriguez, M. (2024). Case-based assessment predicting ICD-11 success. Medical Education Assessment, 22(1), 56-64.
58. Thompson, J. (2023). EHR selection impact on ICD-11 implementation. Healthcare IT Management, 27(3), 145-153.
59. Lee, R. (2024). Training quality correlation with implementation speed. Healthcare Training ROI Quarterly, 8(2), 89-96.
60. Davis, M. (2023). Consultant selection lessons from failed implementation. Healthcare Consulting Review, 14(4), 201-208.
61. World Health Organization. (2024). ICD-11 Implementation Guide Version 2.4. Geneva: WHO Press.
62. Harrison, J. E., & Roberts, K. L. (2023). Post-coordination in practice: Real-world applications. International Classification Quarterly, 45(3), 234-245.
63. American Hospital Association. (2024). ICD-11 Readiness Survey Results. Chicago: AHA Press.
64. Medical Coding Institute. (2023). Common ICD-10 to ICD-11 Mapping Challenges and Solutions. MCI White Paper Series, 2023(7).
65. Healthcare Financial Management Association. (2024). Financial implications of ICD-11 transition. HFMA Special Report, February 2024.
66. National Committee on Vital and Health Statistics. (2023). Recommendations for ICD-11 Implementation in the United States. NCVHS Report to the Secretary, 2023.
67. International Federation of Health Information Management. (2024). Global ICD-11 Implementation Status Report. IFHIM Quarterly Update, Q1 2024.
68. Patel, R., & Martinez, S. (2023). Workflow optimization during ICD-11 transition. Journal of Healthcare Process Improvement, 19(4), 345-356.

69. Anderson, K., et al. (2024). Vendor readiness assessment for ICD-11: Multi-vendor comparative study. Healthcare Technology Assessment, 31(2), 123-134.
70. Thompson, B. (2023). Change management strategies for ICD-11 adoption. Organizational Dynamics in Healthcare, 25(3), 267-278.
71. Robinson, L. (2024). Small practice success with ICD-11: Case studies and strategies. Journal of Ambulatory Care Management, 47(2), 134-142.
72. Healthcare Information and Management Systems Society. (2024). ICD-11 Technology Requirements and Architecture. HIMSS White Paper, 2024(1).
73. Davis, P., & Wilson, K. (2023). Quality assurance frameworks for ICD-11 implementation. Healthcare Quality Journal, 29(4), 189-197.
74. Miller, J. (2024). Training efficacy in ICD-11 transitions: A meta-analysis. Medical Education Research, 18(1), 45-57.
75. Centers for Disease Control and Prevention. (2024). ICD-11 Mortality Implementation Planning. CDC Technical Bulletin, 2024(3).

www.ingramcontent.com/pod-product-compliance
Lightning Source LLC
Chambersburg PA
CBHW070304290326
41930CB00040B/2047